THE NEW ASTRONOMY is a rich kaleidoscope of the finest images of planets, stars, galaxies and the universe. It presents a host of new information, gathered from right across the spectrum: spanning the colourful cosmos from X-rays, through ultraviolet, visible and infrared, and out to radio waves. Nigel Henbest and Michael Marten take us on a journey in which we view the variety of the universe and its contents through every available window.

The first edition of THE NEW ASTRONOMY created a sensation, because it was the first accessible description of modern astronomy to assemble images from so wide a range. For the new edition there are almost 200 entirely new pictures, selected from the Hubble Space Telescope and orbiting X-ray detectors, as well as from the leading ground-based radio and infrared telescopes. The new science includes intriguing images from gravitational lenses, which are natural telescopes created by dark matter around other galaxies, and a full description of the latest images of the background radiation of the universe.

From the nearby planets to, quite literally, the edge of the observable universe, this book is a brilliant synthesis of all that is new in the astronomy of today. Each object described is displayed in a variety of wavelengths. The non-technical text explains the science behind the objects, and it has been substantially re-written for this Second Edition.

the new astronomy

nigel henbest

michael marten

CAMBRIDGE
UNIVERSITY PRESS

Published by the Press Syndicate of the University of Cambridge
The Pitt Building, Trumpington Street, Cambridge CB2 1RP
40 West 20th Street, New York, NY 10011-4211, USA
10 Stamford Road, Oakleigh, Melbourne 3166, Australia

First published 1983
Second edition 1996

Printed in Great Britain at the University Press, Cambridge

A catalogue record for this book is available from the British Library

Library of Congress cataloguing in publication data

Henbest, Nigel
        The new astronomy / Nigel Henbest, Michael Marten. – 2nd ed.
        p.   cm.
        Includes index.
        ISBN 0 521 40324 3 (hc). – ISBN 0 521 40871 7 (pb)
        1. Astronomy.   I. Marten, Michael.   II. Title.
        QB43.2.H463 1996
        520–dc20 95-32981 CIP

ISBN 0 521 40324 3 hardback
ISBN 0 521 40871 7 paperback

Text **Nigel Henbest**
Picture editor **Michael Marten**
Picture research **Caroline Erskine, Gary Evans**
Diagrams **Julian Baum, David Parker**
Design & Art Direction **Richard Adams Associates**
Design **Edwin Belchamber**
Typesetting **Wayzgoose**

Acknowledgements

Special thanks to Heather Couper (expert advice
on galaxies), Gary Evans (digital imaging support)

This book would not have been possible without the many
individuals who located, supplied or processed imagery especially for
us. Particular thanks for their time and trouble to:

Michael A'Hearn, Behram Antia, Rainer Beck, Charles Beichmann,
Ralph Bohlin, Jack Burns, Chris Burrows, Michael Burton,
Dennis di Cicco, Coral Cooksley, Thomas Dame, Emmanuel Davoust,
Antoine Dollfus, José Donas, Martin England, J. V. Feitzinger,
Janice Foster, Ian Gatley, John Gleason, Leon Golub, Martha Hazen,
Sarah Heap, Jeff Hester, Steven Hopkins, N. Hubin, Stan Hunter,
Justin Jonas, G. Kanbach, Kevin Krisciunas, Glen Langston,
Gaylin Laughlin, Charles Lindsey, K.Y. Lo, David Malin,
Michelle Mangum, Craig Markwardt, Michael Mendillo,
Felix Mirabel, Tom Muxlow, Peter Nisenson, Vince O'Connor,
Thomas Prince, Ian Robson, Royal Astronomical Society Library,
Julia Saba, Mike Scarrott, Volker Schönfelder, Nigel Sharp,
Jean-Pierre Sivan, Patricia Smiley, Andy Strong, Keith Strong,
Charlie Telesco, Jonathan Tennyson, Joachim Trümper,
Barry Turnrose, Steve Unger, Steve Unwin, Alan Uomoto,
J. M. van der Hulst, Ray Villard, John Wells, Richard Wielebinski,
Farhad Yusef-Zadeh, Dennis Zaritsky, Giancarlo Zuccotto

contents

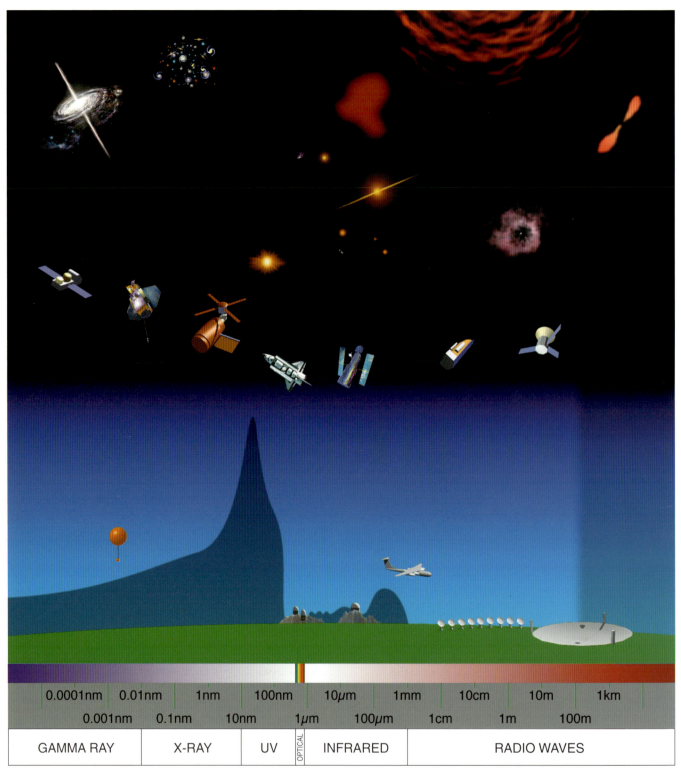

0.0001nm	0.01nm	1nm	100nm	10µm	1mm	10cm	10m	1km
0.001nm	0.1nm	10nm	1µm	100µm	1cm	1m	100m	

GAMMA RAY	X-RAY	UV	OPTICAL	INFRARED	RADIO WAVES

Fig. 1.2 *Astronomers have developed instruments to examine all kinds of radiation from the Universe, ranging in wavelength from the shortest gamma rays (far left) through optical ('rainbow' spectrum) to the longest radio waves (far right). The Earth's atmosphere blocks almost all these wavelengths, however, leaving just two clear 'windows' where astronomers can observe the Universe with telescopes sited at sea-level: the optical window (centre) and the radio window (right), containing the Very Large Array and the Arecibo radio telescope. The long-wavelength side of the radio window is closed by the ionosphere, which reflects long radio waves back to space. Lying between the radio and optical windows is infrared radiation, which is absorbed by water vapour and carbon dioxide in the lower atmosphere: some infrared wavelengths can be observed from high mountain tops or aircraft. To the shorter side of the optical window, ultraviolet (UV) is absorbed by the ozone layer, while gas atoms and molecules in the upper atmosphere block X-rays and gamma rays from space. Astronomers have built satellites to observe almost all these wavelengths unimpeded by the atmosphere (left to right: Compton Gamma Ray Observatory, Rosat, Skylab, Astro, Hubble Space Telescope, Infrared Space Observatory, Cosmic Background Explorer).*

The multiwavelength view has revealed an unsuspected cosmos. At the shortest wavelengths, the sky is dominated by quasars and other 'active galaxies', and by superhot gas in clusters of galaxies. Ultraviolet astronomy has shed new light on the outer layers of our Sun and other stars. Infrared telescopes have laid bare the cool clouds which are the birthplace of stars, many emitting powerful streams of gas. At radio wavelengths, we can observe the remnants of supernova explosions, the vast clouds of electrons and magnetism ejected by distant galaxies and the glow of radiation from the Big Bang.

Fig. 1.3 *Clusters of galaxies contain pools of ultrahot gas that are 'visible' only at X-ray wavelengths. This image combines an optical photograph of the cluster centred on the galaxy NGC 2300 with an X-ray view of the same region of sky (magenta). The pool of X-ray emitting gas is at a temperature of 10 million degrees, but is too tenuous to destroy the galaxies that swim in it. By combining multi-wavelength views, astronomers can learn more than each individually reveals. Optical observations can indicate the gravitational pull of these galaxies: it is not enough to trap the hot gas detected at X-ray wavelengths. So the cluster must contain a huge amount of 'dark matter', which is invisible at any wavelength.*

Visible light is radiation of an 'intermediate' wavelength – about 500 nanometres (one nanometre is a millionth of a millimetre). In everyday terms, the waves are certainly short: over a hundred wavecrests would be needed to span the thickness of this page. The human eye perceives light of different wavelengths as the various colours of the rainbow: red light has the longest waves, around 700 nanometres from crest to crest; and blue-violet is the shortest, with wavelengths of about 400 nanometres. Although these limits to visible light dictate what the human eye can see at the telescope, professional astronomers now use light-detectors that cover a rather wider span. 'Optical astronomy' is generally considered to cover the range from 310 to 1000 nanometres (1 micrometre).

Radiation with shorter wavelengths is invisible to the eye. The range of radiations with wavelengths from 310 down to 10 nanometres is the *ultraviolet*. These are the rays in sunlight which tan our skins; for the astronomer, they are the 'light' from the hottest stars. During its passage through space ultraviolet radiation is imprinted with invaluable information about the tenuous gases between the stars.

At shorter wavelengths than the ultraviolet are the *X-rays*, whose crest-to-crest distance ranges from 10 down to only 0.01 nanometres – the latter is about one-tenth the size of an atom. X-rays from space are the hallmark of superheated gases, at a temperature of over a million degrees. X-ray sources can be the hot gases thrown out by exploding stars, or rings of gas falling down on to a pulsar or a black hole; they can be the hot gases of a quasar explosion, or the huge pools of gas which fill whole clusters of galaxies (Fig. 1.3).

Even shorter are the *gamma* rays – a name which encompasses all radiation whose wavelength is less than 0.01 nanometres. These come from the most action-packed of astronomical objects: from the compact pulsars, and the superexplosions of distant quasars. And gamma rays can spring from nuclear reactions, in regions of space where tremendously fast electrons and protons cannon into atoms in space and provoke nuclear reactions similar to those in the artificial particle accelerators which physicists use to probe the ultimate constituents of matter.

Moving the other way in wavelength, to radiations longer than those of light, we come into the realm of the infrared. These rays have wavelengths between 1 micrometre and 1 millimetre. We think of infrared in everyday life as being heat radiation – the rays from an electric fire, for example. For astronomers, infrared radiation is the signature of the cooler objects in the Universe. An electric fire, at a temperature of a few hundred degrees Celsius, is rather cool on the cosmic scale, where an average star has a temperature of several thousand degrees and some gas clouds have multi-million degree temperatures. Objects at room temperature produce infrared radiation too. We are constantly surrounded by this radiation on Earth, so we do not notice it. But an infrared astronomer can pick up the radiation from a planet which is at everyday temperatures, and even below.

In fact, all objects produce radiation of some kind, and the lower the temperature the longer the wavelength of the resulting radiation. The hot filament of a light bulb naturally glows visible light; an electric fire with shorter wavelength infrared; and our bodies with longer wavelength infrared. Objects which are cooled down until they almost reach the absolute zero of temperature (–273.15°C) emit infrared radiation so long that it technically falls into the region of radio waves. Our view of the sky at different wavelengths is in many ways a portrait of different temperatures; and to the astronomer it makes more sense to measure temperatures upwards from absolute zero – instead of the rather arbitrary Celsius system which is based on the melting point of ice and the boiling point of water at sea-level on our own planet. *Absolute temperatures* thus start at absolute zero: 0 K – where the symbol K stands for degrees Kelvin, named after the physicist Lord Kelvin who first realised the advantages of using absolute temperatures in science. The melting point of ice (0°C) is about 273 K; room temperature roughly 300 K; and the boiling point of water about 373 K – to convert Celsius temperatures into absolute, add 273.15. The absolute scale is actually easier to visualise in one way, because it has no negative temperatures – nothing can be colder than absolute zero.

Infrared astronomers can 'see' cool clouds of dust in space, which are invisible at other wavelengths. These hidden dust clouds are the spawning ground for new stars, and infrared astronomers are privileged to see the first signs of starbirth.

Beyond the region of the infrared lies the last type of radiation. *Radio waves* cover a huge range of wavelengths: technically, they are radiation with a wavelength greater than 1 millimetre. Radio astronomers regularly observe the sky at wavelengths of a few millimetres or a few centimetres, but a few radio telescopes are designed to pick

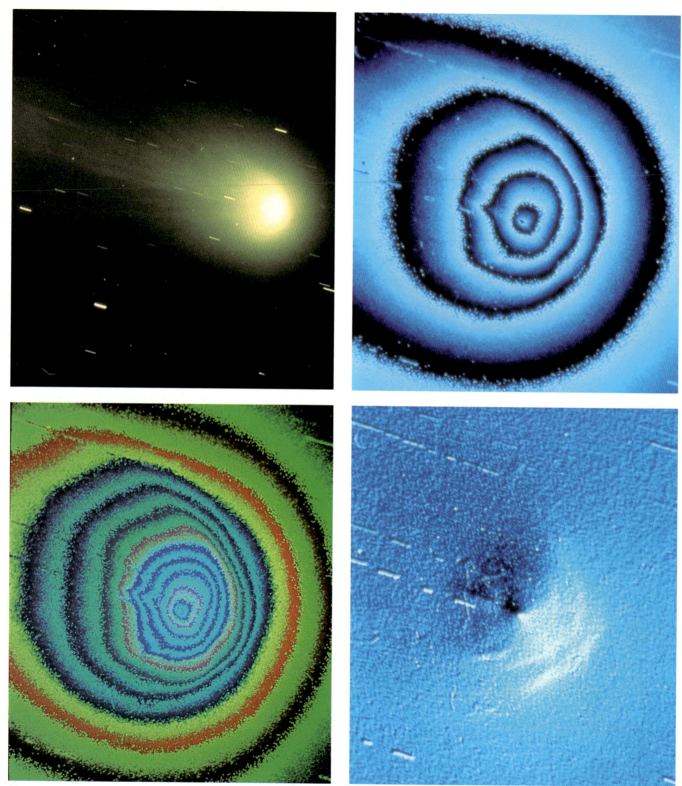

Figs. 1.4-1.7 'New astronomy' techniques of computer processing have been applied here to an old photograph of Halley's Comet, to bring out hidden details. The original photograph, **Fig. 1.4** (top left), was taken in Egypt on 25 May 1910. It shows details in the tail, but relatively little of the structure of the bright head. Four photographs taken that night have been scanned by a small light spot, and their electronic images added by computer to produce the remaining displays (each of which also shows four sections of several star images, trailed as the camera followed the moving comet). In **Fig. 1.5** (top right), brighter levels of the image have been coded by paler shades of blue, up to a level (white) where the coding jumps back to black, and then works up to white again. A succession of such jumps produces the dark and light fringes, like contour lines, surrounding a peak at the brightest, central part of the comet's head. This technique shows clearly a small jet, extending behind (to the left of) the head. **Fig. 1.6** (lower left) shows the same data, but with many more contour levels, and coded in several colours. The extra contours help to indicate the limited extent of the jet. The apparently three-dimensional image, **Fig. 1.7**, has been made by shifting the electronically stored image slightly, then subtracting the shifted image from the original. The comet's head appears as a peak, lit from the lower right, with apparent height indicating brightness. The technique reduces large scale contrasts, and emphasises small scale details. As well as the jet, Fig. 1.7 reveals bright arcs of gas in front of the comet's head (lower right), which are not easily seen in the other representations of the photograph.

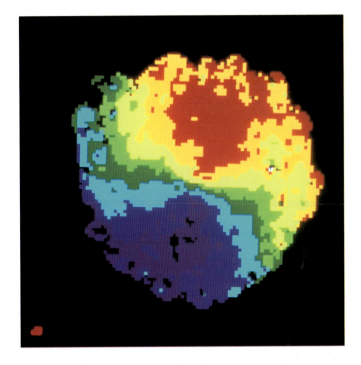

up waves as long as several metres – or even several kilometres. Natural radio broadcasters in the Universe are usually places of unbridled violence, where high-speed electrons – negatively charged subatomic particles – are whirled about in intense magnetic fields.

The telescopes used to observe and detect these radiations are described in the odd-numbered chapters. The results themselves are shown in the even-numbered chapters, with the images obtained at different wavelengths displayed as pictures. These images are routinely stored in a computer, and processed in a wide variety of ways, as discussed in more detail in Chapters 3 and 7.

Using the imaging-processing capabilities of the computer, displayed on the screen, astronomers can employ colour coding to bring out the mass of information in each image. Colour in reality comprises light of different wavelengths. When images are obtained at a wavelength in, for example, the X-ray or radio range, colour has no real meaning, and we can use colour coding in a variety of new and exciting ways (Figs. 1.4–1.7). Colour is employed in three main ways in this book.

Observations at different wavelengths, combined into a single image, can be colour coded according to the wavelength. This technique is often used in infrared astronomy, where astronomers make observations of the same object at different wavelengths to reveal differences in temperature. Blue generally represents the shortest-wavelength view, green an intermediate wavelength and red the image at the longest wavelength.

Alternatively, colour coding can show velocity. The speed of a gas cloud is revealed in observations of a single spectral line from atoms or molecules of gas, usually obtained at radio wavelengths. By convention, gas coming toward us is shown in blue, and the colours shade through to red for gas moving away (Fig. 1.8).

But colour coding is most commonly used to show intensity. This technique is widely used at all wavelengths – including optical, when black-and white-images are processed by computer. We can assign different colours, chosen at will, to the various levels of brightness in the picture. The technique is now pretty familiar not just to astronomers but to anyone who watches a pop video!

The results are not only picturesque. They overcome one of the problems of photographic representation, that it is impossible to show details in both the faintest part of a galaxy or nebula, and the brightest – which may be over a thousand times more brilliant. If the former are shown in a photograph, the bright regions are 'burnt out'; while a short exposure to reveal details in the bright regions would not show the faint parts at all. Intensity colour coding shows details of both bright and faint regions simultaneously, in different colours.

The new images often resemble works of art, with the Universe's natural artistry aided by the imagination of the astronomer at the computer. But the astronomer's main task is not to capture the unseen beauty of space, but to use this new information to help understand the structure and scale of the Universe, and how it is changing as time goes by.

In relation to the modern view of the Universe, our planet Earth is a mere speck along with the other eight planets circling the Sun. The planets range in size from Pluto, whose diameter is one-sixth of Earth's, to Jupiter with a diameter of eleven Earths. Saturn is almost as large as Jupiter, and these 'giants' of our Solar System are orbited by 18 and 16 moons, respectively.

Despite the size of the giant planets, they lie so far away that astronomers find it difficult to see details on them. The Earth orbits the Sun at an average distance of 150 million kilometres, but Jupiter's orbit is over five times larger, and Saturn's nearly twice as big again. So these worlds appear very small in our sky: although our unaided eyes can see them shining brightly in reflected sunlight, they seem to be no more than points of light. It needs a telescope to show their globes, and Saturn's encircling girdle of rings.

Astronomers measure the apparent size of objects in the sky in terms of *degrees of angle* (°), and their subdivisions *arcminutes* and *arcseconds*: there are 60 arcminutes to one degree, and 60 arcseconds to one arcminute. This traditional system is undoubtedly cumbersome, but it becomes easier to understand if we take some examples. The entire sky is 180° across, from one horizon up to the zenith and down to the opposite horizon. Most of the traditional constellation patterns are around 20° across: the figure of the 'hunter' Orion (Fig. 4.2), for example, stands 15° tall. The Moon is surprisingly small. Because it shines so brightly, and appears large when it is near the horizon, the Moon looks as though it should cover quite an area of sky; but in fact it is only half a degree across (30 arcminutes). It would take over three hundred Moons, put side-by-side, to stretch across the sky.

Our eyes cannot see details any finer than one or two arcminutes (about $\frac{1}{20}$ the size of the Moon), and the planets

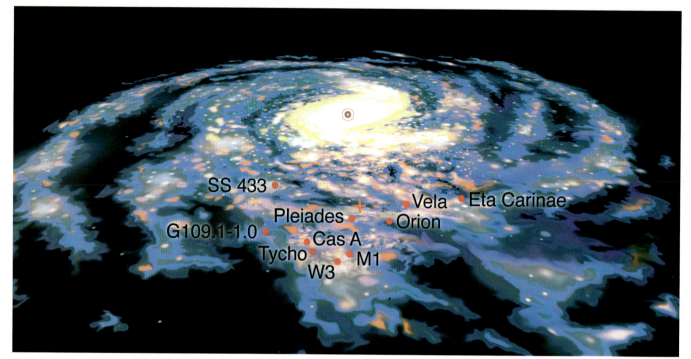

Fig. 1.9 *On the outskirts of the Milky Way Galaxy lie the Sun and Solar System (marked as +), which feature in Chapter 2. Other inhabitants of the Galaxy appear in later even-numbered chapters. Chapter 4 includes the starbirth regions W3 and Orion, the powerful young star Eta Carinae and the Pleiades (a clutch of newly-born stars). Supernova remnants, the gaseous remains of long-expired stars, are featured in Chapter 6: G109.1-1.0, SS 433, Cassiopeia A, Tycho's supernova remnant, the Crab Nebula (M1) and the Vela supernova remnant. Chapter 8 covers the Milky Way itself, and the galactic centre (circled), some 25 000 light years from the Sun.*

all appear smaller than this. Even when the Earth is closest to Jupiter, the giant planet appears just under an arcminute across and Saturn only 20 arcseconds in size. Even the largest optical telescopes on Earth can see detail only as fine as an arcsecond or so, whatever magnification is used, because the images are blurred by Earth's atmosphere. So, ironically enough, astronomers have problems in examining even our nearest neighbours in space. These worlds are close enough, though, to be reached by unmanned spaceprobes, and the last three decades have seen their details revealed by dozens of these craft – from Mariner 10 which visited scorched Mercury, closest to the Sun, to the two Voyagers which have photographed in amazing detail four giant planets: vast Jupiter, ringed Saturn, tipped-up Uranus and remote Neptune.

On the everyday scale, Neptune is a very distant world. Lying some 4500 million kilometres away, it is so far off that the radio signals from Voyager 2 – travelling at the speed of light – took over four hours to reach the Earth. Until 1999, Neptune is the outermost planet, because Pluto follows an oval path that has currently brought it closer to the Sun than Neptune.

But even Neptune's distance shrinks into insignificance once we look out to the stars. The nearest star, Proxima Centauri (part of the Alpha Centauri star triplet), lies a quarter of a million times farther away than Neptune and Pluto – over 40 million million kilometres. Such distances are just too large to comprehend, and the figures themselves become unwieldy in size.

Astronomers cope with star distances by discarding the kilometre as their standard length. A more convenient unit is the distance that a beam of light (or any other radiation) travels in one year. This standard length, the *light year*, works out to just under ten million million kilometres. So

Proxima Centauri is about 4 1/4 light years away. With this new unit, star distances become more comprehensible. We can compare star distances, and construct a scale model in our minds, even if the distances themselves are too large for the human mind to encompass.

The bright star Sirius lies about twice as far away as Proxima Centauri, at 8.6 light years. It too is nearby on the cosmic scale; its neighbours as seen in the sky, Betelgeuse and Rigel in Orion, are far more distant, lying at 310 and 910 light years. They only appear bright in our skies because they are truly brilliant stars, each shining as brightly as thousands of Suns.

Optical astronomers have a rather archaic system for describing the apparent brightness of stars and other celestial objects as seen from the Earth. In this *magnitude* system, the faintest stars visible to the naked eye are of magnitude 6; while brighter stars have *smaller* magnitude numbers. A star of magnitude 1 – like Betelgeuse – is a hundred times brighter than a magnitude 6 star; the brightest star, Sirius, is ten times brighter still, and it has a negative magnitude: –1.4. Going brighter still, the Sun has a magnitude of –26.7!

Our Sun is, in fact, a typical star – middle-weight, middle-aged – born some 4600 million years ago. It appears exceptionally bright simply because it is near to us, and it is special to us on Earth because it supplies us with bountiful light and heat. The Sun is special to astronomers too, because its proximity means that we can study an average star in close-up detail. Conversely, investigations of other stars which are younger and older can tell us of the Sun's past, and of its likely future.

Starbirth occurs because space between the stars is not entirely empty. There is tenuous *interstellar gas*, composed mainly of hydrogen, with tiny flecks of dark solid dust

Fig. 1.10 *Our Galaxy, the Milky Way (MW), and its companion the Large Magellanic Cloud (LMC) are scrutinised in Chapter 8. With the Andromeda Galaxy (M31) and the smaller spiral M33 they comprise the major galaxies of our Local Group. Two neighbouring groups include M51, M81, M82 and M101. Chapter 10 covers these six relatively normal galaxies. Further afield lie more exotic cosmic beasts, described in Chapter 12. They include the radio galaxies Centaurus A, M87 (in the Virgo Cluster), NGC 1275 (in the Perseus Cluster) and Cygnus A, along with the quasar 3C 273 and the gravitational mirage seen as the 'twin quasar' (marked as 0957+561).*

mixed in. In places, this gas and dust is compressed into sombre dark clouds. Within these, the interstellar matter is compressed by its own gravity into gradually-shrinking spheres, which heat up until they burst into radiance as new stars. The brilliant young stars light up the surrounding tatters of gas as a glowing *nebula*, like the famous Orion Nebula (Figs. 4.10–4.17), a mass of seething fluorescent gases lying some 1600 light years from us.

Eventually, all stars die. They lose their outer gases, either in gentle cosmic 'smoke rings' or in violent supernova explosions; while their cores collapse to form tiny, compact – and very strange – objects. These cores weigh roughly as much as the Sun, but they can shrink down to the size of a planet to form a *white dwarf* star; even smaller as a *neutron star* or pulsar; or even collapse completely to form a *black hole*, whose gravitational field is so powerful that no radiation can escape from it.

Stars are grouped into huge star-islands, or *galaxies*. Our Sun (and the other stars and nebulae mentioned so far) is a member of the *Milky Way Galaxy* (Fig. 1.9), a collection of around 200 000 million stars – along with nebulae and collapsed star-corpses. The stars are arranged into a huge disc, some 100 000 light years across, all orbiting around the Milky Way's centre where the disc of stars is thicker. Our Sun, with its family of planets, lies about two-thirds of the way to the edge of the Galaxy.

The Milky Way's nearest neighbours are two smaller galaxies, the Large Magellanic Cloud and the Small Magellanic Cloud. They can only be seen from the Earth's southern hemisphere, and the two Clouds were first described by the Portuguese navigator Ferdinand Magellan as he circumnavigated the Earth in 1521. Farther away lies the great Andromeda Galaxy, dimly visible from the Earth's northern hemisphere in autumn and winter months. At

2¼ million light years distance, it is the farthest object we can see with the naked eye.

But telescopes can reveal many other, much more remote galaxies (Fig. 1.10), their intrinsic brilliance dimmed by their enormous distances. While the Andromeda Galaxy has a magnitude of five, rather brighter than the naked eye limit, modern telescopes, coupled with sensitive electronic detectors, can 'see' galaxies almost a billion times fainter, around magnitude 30. Such galaxies lie over 10 000 million light years away from us. Within this vast region of space, the latest telescopes can pick out thousands of millions of galaxies. Many of them are clumped together into huge clusters, each cluster stretching over millions of light years.

Amongst this multitude of galaxies, a few display an intensely bright core, the site of a stupendous explosion – an explosion which can be a thousand times brighter than the galaxy itself. If the galaxy is extremely remote we cannot see anything but this central explosion – and astronomers have called such objects *quasars*. These exploding galaxy cores are so bright that astronomers can see them farther away than any other object in the Universe. The most distant quasars lie some 12 000 million light years away from us. The quasar explosions generate huge quantities of radio waves and X-rays too, and some of the many radio sources and X-ray sources discovered recently are probably quasars even more remote.

Centuries of observations with the eye and optical telescope have revealed the framework of the Universe, but their information on the planets, stars and galaxies has turned out to be only superficial. The story of the life and death of stars, of galaxies and of the Universe itself has only become apparent in recent years, with the advent of the new astronomy.

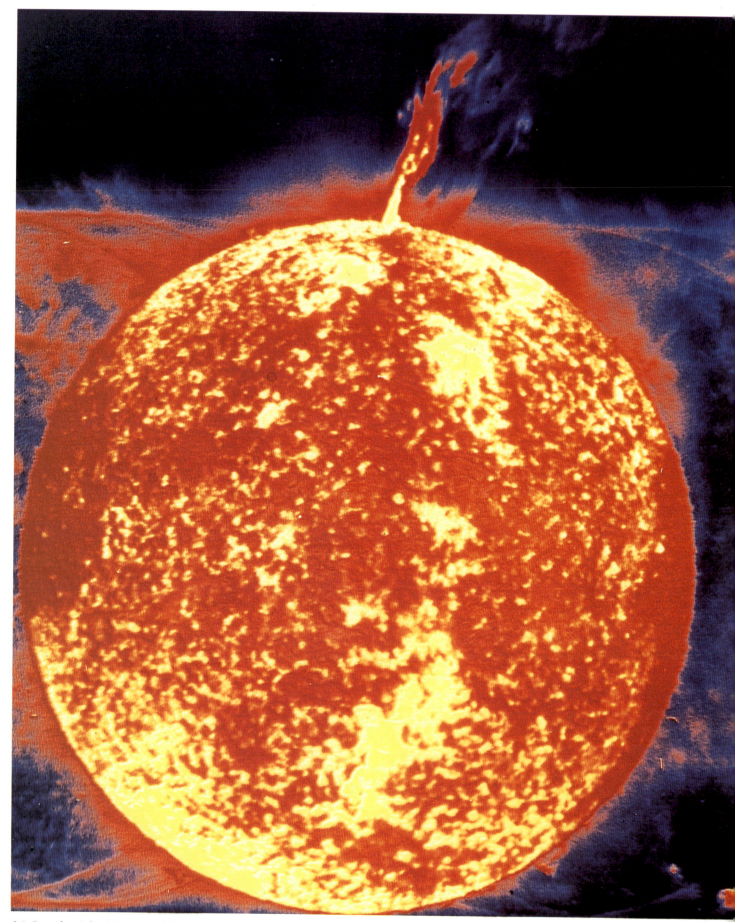

2.1 *Sun, ultraviolet, 28–31 nm, Skylab extreme ultraviolet spectroheliograph*

2 solar system

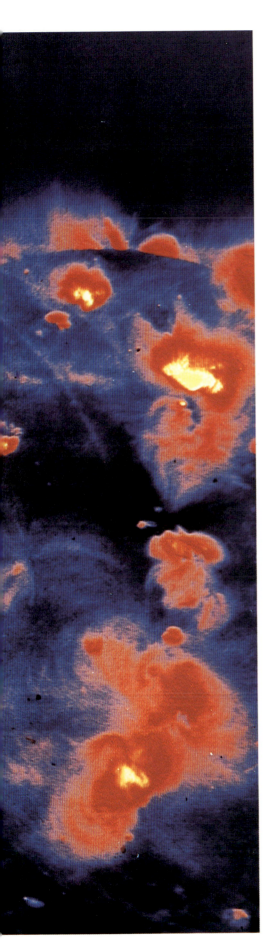

THE SOLAR SYSTEM is dominated by the Sun – and by sunlight. The Sun contains 99.9 per cent of the mass of the system, so its gravitation controls the motions of the planets and the minor members of the system, like the asteroids and comets. The Sun is also the only important source of energy in the Solar System. Its central nuclear reactions are over a hundred million times more powerful than the interior heat sources of the planets combined. Most of this power is emitted from its surface as visible light.

Sunlight reflected from the planets makes them prominent in our night skies. Venus can shine ten times more brilliantly than the brightest star, while Mercury, Mars, Jupiter and Saturn all rival or exceed the half-dozen brightest stars. The appearance of the planets in visible light has become familiar from telescopic views and, more recently, from spaceprobe pictures.

But in solar system astronomy – as in the study of the Universe beyond – other wavelengths of radiation can reveal new aspects of familiar objects. The Sun itself is no exception. Its powerful output of visible light produces a uniform glare from the whole disc which makes it difficult to study details of its surface or atmosphere.

Structure in the surface is best seen by isolating light of particular wavelengths, those emitted and absorbed by common atoms like hydrogen. Ultraviolet views show details of the lower atmosphere and X-rays the upper atmosphere – both virtually invisible to optical astronomers except during rare solar eclipses. These wavelengths, and radio waves, reveal that the Sun's surface and atmosphere teem with activity, and occasionally suffer powerful outbursts – unseen by optical astronomers – which can affect the Earth.

The long-term study of ultraviolet radiation and X-rays has required satellite observations above the Earth's absorbing atmosphere, and the manned space station Skylab was equipped with a specialised solar observatory. **Fig. 2.1** was taken at wavelengths around the middle of Skylab's range, in extreme ultraviolet radiation. The Sun's surface is not visible here: the picture instead contains several overlapping images of its lower and upper atmosphere. In Fig. 2.1, the Skylab instrument has spread the Sun's radiation out into a spectrum, with the wavelength increasing from right to left. This radiation comes from atoms in the atmosphere which radiate only at particular wavelengths – forming a complete image of the Sun at each wavelength of emission.

Within each image, dim regions are coded blue, and brighter parts red and yellow. (North is to the right in each image.) The full disc at the left is seen at a wavelength of 30.4 nanometres, in the radiation from helium atoms which have lost one electron. These occur in the lower atmosphere, and so the mottled disc shows the clumpy distribution of gas just above the visible surface – along with a huge prominence of gas projecting half a million kilometres – almost 50 Earth-diameters. The right-hand image, at 28.5 nanometres, comes from iron atoms missing 14 of their 26 electrons, because they lie in very hot gas clouds. The image reveals these isolated clouds in the outer atmosphere above the Sun's active regions, but not the cooler gas of the Sun's surface, and of the lower atmosphere and its prominence.

The planets have not been as intensely studied at non-optical wavelengths as the Sun, and the results have been overshadowed by detailed optical photographs from spacecraft. Ultraviolet and X-ray observations – from spaceprobes or from observatories in Earth orbit – can, however, reveal activity in planetary atmospheres which is invisible in their reflected sunlight. Astronomers have gleaned even more information from the longer wavelengths – partly because they can be observed from ground-based observatories. All the planets emit infrared 'heat radiation' simply because they are warmed by the Sun. Infrared astronomy also opens our view to the depths of the atmospheres of the large gaseous planets Jupiter and Saturn, which are warmer inside than at the visible cloud tops.

Radio astronomy plays two roles. Planets emit radio waves from their warm surfaces, or by *synchrotron* emission from electrons trapped in their magnetic fields. The latter reveals the extent, shape and strength of the field. In planetary *radar*, a beam of artificially generated radio waves is sent towards a planet, and the faint returning 'echo' is detected. This technique provides the most accurate distances to the planets, and can show details on the surfaces of solid planets. Since radio waves can penetrate clouds, radar astronomy has provided the first global views of the nearest planet, cloud-shrouded Venus.

The Sun

'A STAR ON our doorstep' is the astronomer's view of the Sun. Our local star is quite an average specimen in terms of its mass, size, temperature, and position in our Milky Way Galaxy (Fig. 1.9); it is halfway through its life span of some 10 000 million years. It has the advantage of being the one star we can study in great detail.

But in everyday terms, the Sun is colossal and awe-inspiring. It is a ball of hot gas, 1.4 million kilometres in diameter and containing the mass of 330 000 Earths. It consists mainly of hydrogen and helium, with a trace of the other elements, and at its core the temperature is so high – 14 million K – that nuclear fusion reactions convert hydrogen into helium. The reactions are similar to those in a hydrogen bomb, but self-controlled to produce a steady output of energy – at the rate of a million million hydrogen bombs every second. Since energy is equivalent to mass, every second the Sun loses four million tonnes of its matter as radiation. This energy starts at the Sun's core in the form of X-rays and gamma rays. The Sun's interior becomes cooler away from its centre, and the radiation's wavelength increases accordingly as it percolates outwards, until it emerges from the 5800 K surface in the form of 'light and heat'. Just over half the Sun's output is visible light, and almost all the rest is infrared with wavelengths less than two micrometres.

As seen optically (**Fig. 2.2**), the Sun is a uniformly shining sphere, with radiation emerging equally in all directions. The surface or *photosphere* seen in photographs like Fig. 2.2 is not a true boundary like the surface of a planet. It is the outer edge of the region of opaque gas, which in fact merges into the transparent gas beyond (the Sun's 'atmosphere'). The density of the Sun's gas decreases very abruptly in the photosphere, so the Sun appears to have a surprisingly sharp edge. The photosphere is about 500 kilometres thick, and near the centre of the Sun's disc we see down deeper into its hotter and brighter lower regions than we can towards the Sun's edge, where the radiation has to traverse more gas. Hence the Sun's disc appears brighter at the centre than at the edges.

The obvious features of Fig. 2.2 are the black blemishes called sunspots, the largest in Fig. 2.2 being as big as the Earth. The number of sunspots varies considerably, and Fig. 2.2 was taken when the Sun displayed an unusually large number. Sunspots are one manifestation of solar *active regions*, regions where strong magnetic fields break through the photosphere into the atmosphere. At optical wavelengths, the active regions appear in a negative role, as their

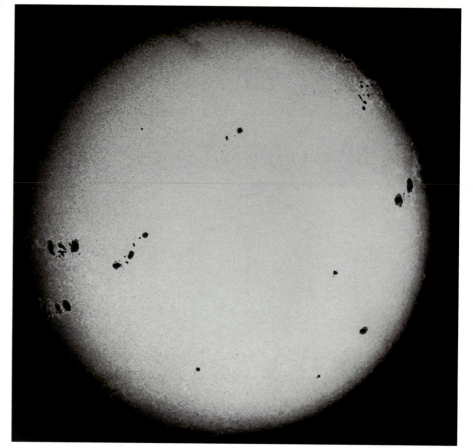

2.2 *Optical (15 September 1957), 46 m tower telescope, Mount Wilson Observatory*

2.3 *Radio (26 September 1980), 20 cm, Very Large Array*

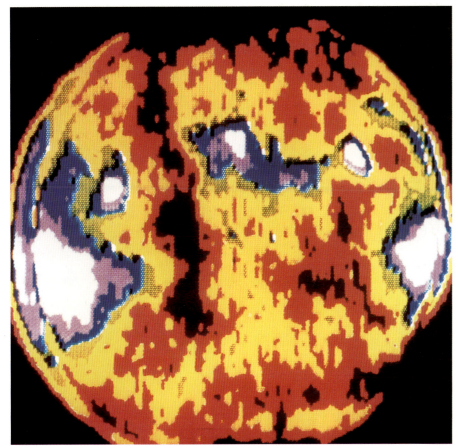

2.4 *Ultraviolet (30 May 1973), 17–55 nm, Skylab extreme ultraviolet monitor*

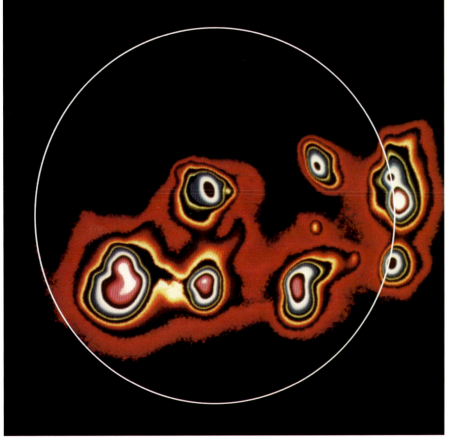

2.5 *X-ray (6 September 1973), 0.6–3.3 nm, Skylab X-ray telescope*

magnetism blocks the Sun's general output of radiation to produce dark sunspots – although there are also slightly brighter areas, *plages*, surrounding the spots. But at most other wavelengths, the active regions are the brightest parts of the Sun.

The radio view (**Fig. 2.3**) shows what looks almost like a negative of the optical Sun. Faint regions are here colour coded blue, with brighter regions green, yellow, red and white. The red and white regions are clouds of hot gas trapped in the magnetic field above active regions; a simultaneous optical photograph would reveal sunspots below each. Fig. 2.3 shows very clearly that active regions occur on the same latitude on the Sun, north and south of its equator. The fainter (blue) regions are more diffuse hot clouds lying above the photosphere. Because we can see these clouds round the Sun's edge, the Sun seems larger when observed in the radio.

Radiation of very short wavelengths comes from extremely hot gas, and the photosphere is so 'cool' that it appears black in ultraviolet and X-ray pictures. The photosphere and the lowest layer of the atmosphere, the *chromosphere*, are at about 5800 K; but at the top edge of the chromosphere the temperature rises abruptly through a thin *transition layer* to reach 1 million K or more in the outer atmosphere, the *corona*. The gases of the transition layer shine in the ultraviolet and the shorter extreme ultraviolet (EUV) radiation, while the corona is brilliant in X-rays.

Fig 2.4 shows our local star in the radiation from gas in the transition layer at a temperature of around 100 000 K. Less intense regions are coded red, and brighter parts yellow, green, blue, purple and white. The transition region does not cover the dark photosphere uniformly. Running down from the North Pole, there is a long dark rift, the lowest layer of a coronal hole – a low-density region of the Sun's atmosphere (Fig. 2.14). The Sun's poles (top and bottom) also have no overlying transition layer. The magnetic fields which generally confine the Sun's atmosphere are here directed straight out into space – like the lines of force from the ends of a bar magnet – and the gas escapes freely. The gas is most confined by the strong fields over active regions, which therefore appear brightest in Fig 2.4.

Farther out, the fields of the active regions trap the hotter, X-ray emitting gas of the corona (**Fig. 2.5**). In this contoured map, regions of successively greater brightness are enclosed in red, yellow, green, blue, pink and white (with black between the colours); the edge of the photosphere is marked by the thin white circle. This gas is generally at 2 to 3 million K, but the intense white regions are at 5 million K, almost a thousand times hotter than the photosphere below.

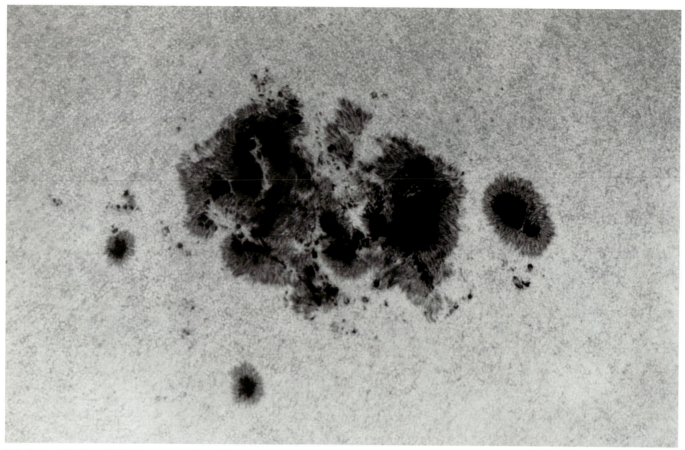

2.6 *Optical (17 May 1951), 46 m tower telescope, Mount Wilson Observatory*

Sunspots are usually studied in visible light, at the wavelengths at which the photosphere and details within it are most easily discerned. The number of sunspots comes and goes in a cycle of 11 years. New spots first appear at high latitudes, and later spots in the cycle nearer to the Sun's equator. A small spot will last a few weeks, while the largest may take several months to grow and then shrink away again.

Spots generally come in pairs, or in complex groups of spots. The straight-forward optical photograph (**Fig. 2.6**) shows a typical large complex – several times bigger than the Earth. Each spot has a dark core, the *umbra*, and a paler outer *penumbra* composed of alternating bright and dark filaments of gas. (The small-scale mottling outside the spot group is the pattern caused by energy 'bubbling up' through the photospheric gas.)

Fig. 2.7 reveals details inside a sunspot's umbra by the technique of speckle interferometry, which reduces blurring by the Earth's atmosphere. In each image, the brightest regions are coded pale pink, and darker regions red, yellow, blue and black. The top frame shows a very small spot, about 4 arcseconds in apparent size (3000 kilometres in diameter) comparable to the very smallest spots visible in Fig. 2.6. An electronic detector, which responds to a wide range of brightness from the 'black' umbra to the photosphere, took a series of

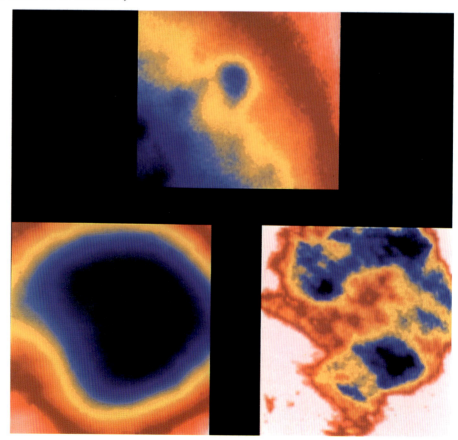

2.7 *Optical (17 July 1981), 405–415 nm, speckle processed, CID camera, McMath Solar Telescope, Kitt Peak*

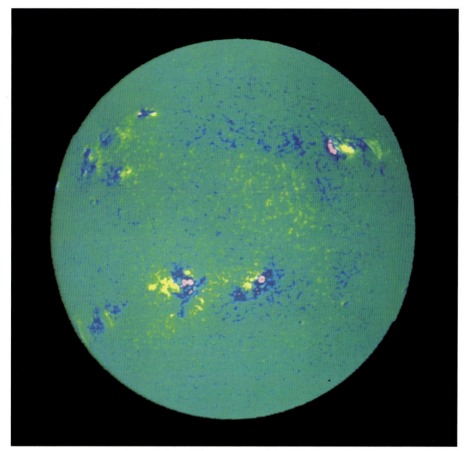

2.8 *Magnetogram, from optical data (3 January 1978), McMath Solar Telescope, Kitt Peak*

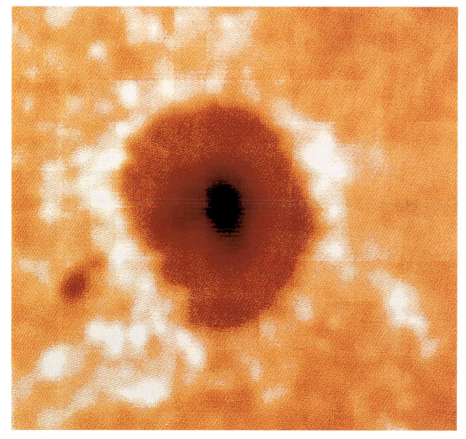

2.9 *Infrared (29 September 1992), 12.4 μm, McMath Solar Telescope, Kitt Peak*

46 very short exposures to 'freeze' the atmosphere's blurring. One of these appears as the magnified image at lower left in Fig. 2.7. The exposures were added to produce the final image of the spot seen at lower right. Clearly, the interior of a sunspot is not uniformly dark: this particular spot is darker around the edges, and slightly less dark in the centre. There is also smaller structure, down to a size of 0.3 arcseconds (200 kilometres), including striations stretching to the top left.

Sunspots occur where powerful magnetic fields break through the photosphere. The Sun's general magnetic field is similar in strength to the Earth's, but the field in sunspots is over a thousand times stronger. The field is concentrated in many smaller ropes, seen as they spread out from the umbra as the filaments in a spot's penumbra. The fine details in Fig. 2.7 may indicate that the individual ropes are narrower and contain a more concentrated field than had previously been thought.

The magnetic field in sunspots affects the spectral lines from gas in them. Optical spectral studies not only show the field strength, but also the magnetic polarity – whether it is a 'north pole' or a 'south pole'. The magnetogram (**Fig. 2.8**) shows the magnetism of all the spots on the Sun, colour coded yellow for 'north pole' and blue and pink for 'south pole', on a turquoise disc indicating the photosphere. Each pair of sunspots clearly includes one spot of each polarity, with opposite orientation in the two hemispheres. The magnetic fields run between them like the lines of force of a horseshoe magnet.

The patches of strong magnetism suppress the currents of hot gas carrying energy upwards from the Sun's interior. The result is a small cool region, radiating less energy than its surroundings and thus appearing as a dark sunspot. (But the darkness of a sunspot is only relative: viewed on its own, a large sunspot would be as bright as the Full Moon.)

The infrared image (**Fig. 2.9**) reveals directly the temperatures in and around a sunspot the size of the Earth, with a precision of 10°C. White denotes the hottest regions, with progressive cooler parts yellow, orange, red and black. The photosphere, at 5500°C, appears yellow, while the core of the sunspot (black) is at only 4500°C.

The filaments of the penumbra appear faintly as orange stripes within the spot, and the infrared image shows clearly how the spot becomes progressively cooler towards the centre. The temperature structure of a sunspot is much more symmetrical than its appearance at visible wavelengths.

The white rim around the spot in Fig. 2.9 indicates that the magnetic field is not a totally effective lid on the Sun's rising energy. Some of the trapped heat is managing to escape sideways and well up around the edge of the sunspot.

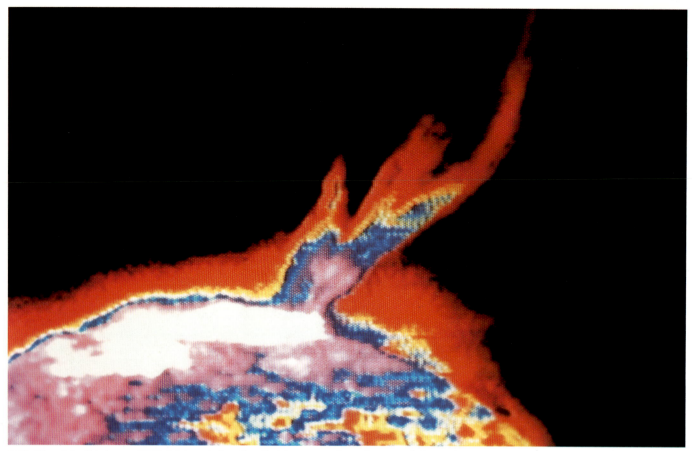

2.10 *Ultraviolet (21 August 1973), 30.4 nm helium line, Skylab extreme ultraviolet spectroheliograph*

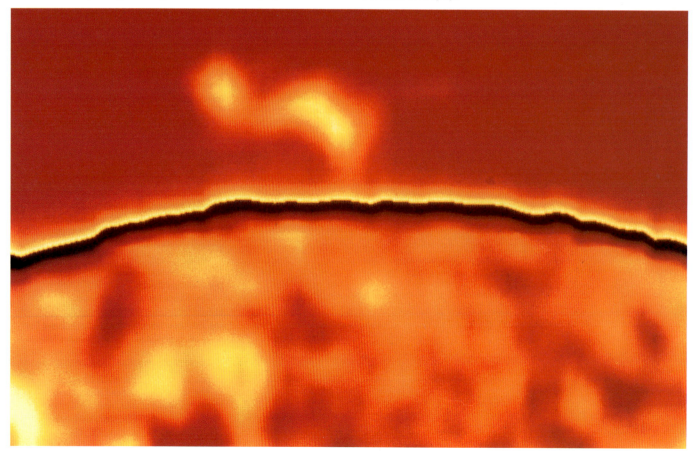

2.11 *Radio (10 July 1991), 1.3 mm, James Clerk Maxwell Telescope*

'Active regions' on the Sun not only experience the gradual build-up of magnetic field which gives rise to sunspots, prominences and loops in the corona; they also produce powerful outbursts and explosions. These great eruptions – virtually invisible at optical wavelengths – send gas and fast electrons streaming through the Solar System. When they reach Earth, the particles produce aurorae in the atmosphere and radio blackouts as they disrupt the ionosphere; and they present a radiation danger to astronauts above the protection of Earth's atmosphere.

Eruptive prominences are one type of solar outburst. They begin as normal prominences, dense curtains of gas at a relatively cool temperature (about 10 000 K) hanging high up in the much hotter corona. Prominences are supported by the magnetic fields that stretch from one sunspot or sunspot group to another. Occasionally, this magnetic field will change its shape and structure quite markedly in just a few minutes. As it twists and tears itself away from the Sun at one end, the magnetic field catapults the prominence's gas out into space. The quiescent prominence becomes an eruptive prominence.

Fig. 2.10 is a view of an eruptive prominence, taken in the light from helium which reveals the chromosphere (seen also in Fig. 2.1). The faintest regions here are coded red, and brighter regions yellow, blue, lilac and white (North is to the right). The white area is an active region associated with a sunspot group, and the magnetic fields have flung a prominence out into space as the multi-stranded jet of matter stretching out half a million kilometres.

But not all prominences erupt into space. A *quiescent prominence* hangs above the Sun's surface, sometimes for weeks or even months. **Fig. 2.11** shows a quiescent prominence imaged at very short radio wavelengths known as millimetre waves. (Indeed, it was the first millimetre-wave observation of a solar prominence, obtained in Hawaii as the astronomers were waiting to observe how the Sun's millimetre wave emission changed during a total solar eclipse that swept over the observatory just 23 hours later.)

The curved dark line is the Sun's edge: the patchy emission beneath comes from hot gas just above the photosphere. Above hangs the S-shaped prominence, appearing yellow in this colour-coding.

The prominence in Fig 2.11 is 75 000 kilometres – six Earth diameters – above the Sun's photosphere. The millimetre-wave observations show that the gas in the prominence is at a temperature of 5000 K, and has a density of 10 000 million atoms per cubic centimetre. Compared to the Earth's atmosphere, this is almost a vacuum; but it means that the prominence is much denser than the tenuous corona

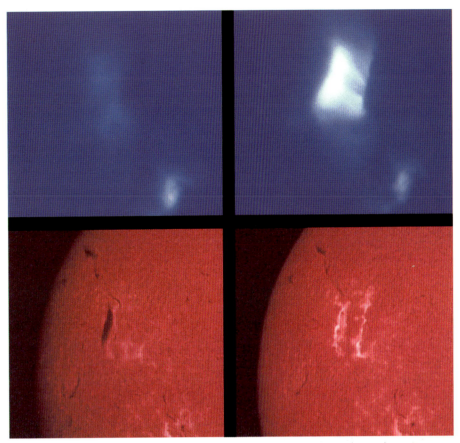

2.12 *X-ray (29 July 1973), 0.2–1.7 nm (blue), Skylab X-ray telescope; optical (29 July 1973), 656 nm hydrogen line (red), ground-based solar telescope*

around it. The coronal gases cannot buoy up the dense filament any more than a brick can float in water, and by rights the prominence ought to crash onto the Sun's surface. The only force that can be supporting this huge prominence is magnetism, and this observation is strong evidence of the power of the magnetic fields generated in sunspots.

The power of the Sun's magnetic fields is shown most dramatically by solar *flares*. These outbursts generate a huge amount of radiation, and accelerate electrons up to high speeds, but they usually eject relatively little gas into space. A flare starts as a small very intense source of radiation and spreads along the lines of magnetic field in a matter of minutes, like the flame from a match tossed into a pool of petrol, and sets a region larger than the Earth shining with the brilliance of a million hydrogen bombs. Despite this output of radiation, a flare cannot match the photosphere's output of light, and flares are usually visible at optical wavelengths only by isolating the light from a single type of atom which is set glowing in the flare. But at other wavelengths, a flare totally dominates the Sun.

Fig. 2.12 compares views taken in X-rays (blue) with optical pictures in the spectral line emitted by hydrogen (red). Before the flare (left-hand images), the small active region at the centre is quiescent. The

optical image shows some plage regions, and a prominence in silhouette, but this region has little in the way of hot gas above it to produce X-rays.

Suddenly, and unpredictably, a flare begins (right-hand images). In a few minutes, it has spread and lit up the plage regions, and the prominence. Although this brightening is visible in the optical (red) photograph, it is not very obvious. The flare has, however, heated up the gas in the prominence until it shines brilliantly at X-ray wavelengths (blue). In a flare like this, the coronal gas can reach a temperature of 20 million K – even hotter than the Sun's core.

The flare's power comes from the energy pent up in the magnetic fields of sunspot regions. When magnetic field loops of opposite polarity approach one another too closely, they can annihilate. This magnetic field reconnection is very similar to an electrical short-circuit, and like a short-circuit it produces a sudden flash of energy – the solar flare. Once a reconnection is made, it can spread rapidly along the divide between two magnetic field regions – often marked by a prominence – liberating energy all along the way. Magnetograms show that an active region's magnetic field structure is simplified after the trauma of a flare has short-circuited the kinks in its complex magnetic pattern.

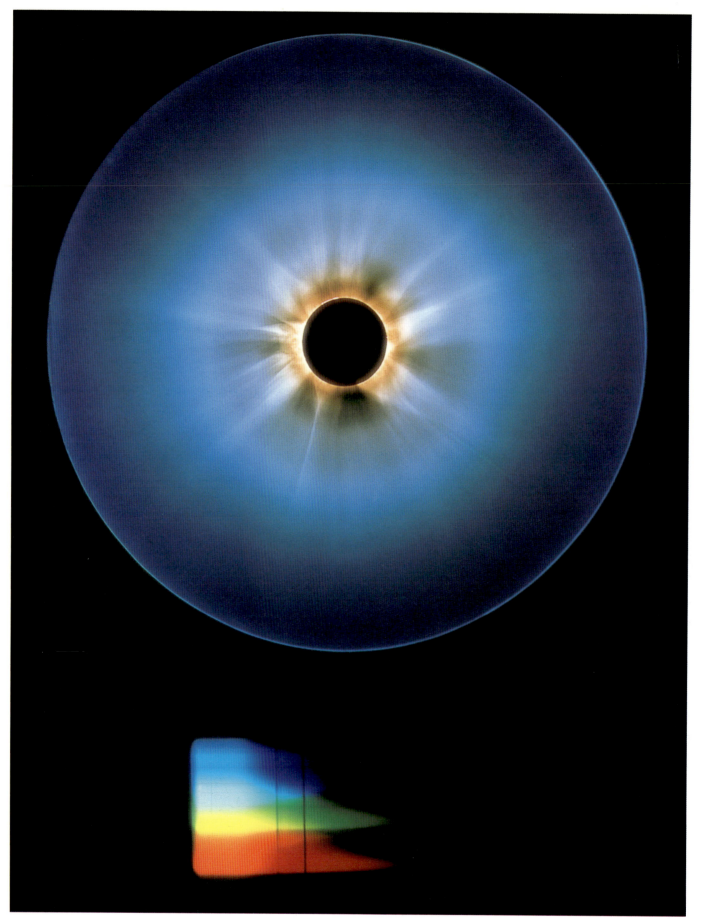

2.13 *Optical, during eclipse (16 February 1980), true colour, eclipse camera with radial gradient filter*

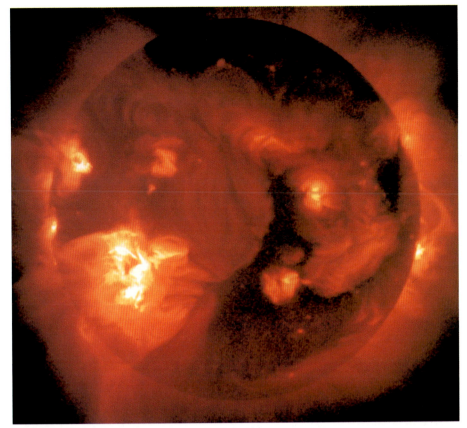

2.14 *X-ray (25 October 1991), 0.4–6 nm, Yohkoh*

2.15 *Composite (30 June 1983): optical, during total solar eclipse; X-ray, Skylab X-ray telescope*

in the magnetic fields of active regions, extending far into space. The 'rainbow' (bottom) is a crude spectrum of the corona, obtained simultaneously. Note the strong green light from iron atoms, and red light from hydrogen.

Eclipses occur only rarely and last but a few minutes. An X-ray view of the Sun (**Fig. 2.14**) reveals the hot coronal gas – which naturally emits these wavelengths – without the need to hide the photosphere, which is dark in X-rays. Hence we can see the corona across the Sun's face. The corona is extremely patchy, the hot gas being only noticeably present in the active regions, where it is concentrated into the loops of magnetic field joining sunspots.

The long dark gap is a *coronal hole*. Here the magnetism of the Sun's surface does not form closed loops, but extends straight out so hot gas is not trapped near the Sun.

A composite image (**Fig. 2.15**) shows just how the X-ray hot-spots are related to the extended corona. It combines an X-ray picture from Skylab with a simultaneous photograph of the Sun in total eclipse.

The helmets clearly lie directly above the X-ray hot-spots. The gases forming both these structures are held in place by the magnetic fields of the active regions, anchored in the sunspots way below. Astronomers believe that the Sun's magnetic field is responsible for heating the coronal gases to their multi-million degree temperatures, but it is still not clear quite how this happens.

Despite the spectacular look of the streamers extending from the helmets, the action is to be found in the gaps between them. Here gas from the coronal holes is escaping the Sun's restraining influence and speeding into space at a speed of hundreds of kilometres per second. This *solar wind* sweeps past the Earth about five days after leaving the Sun, when the high-speed particles impinging on the top of the atmosphere can cause a display of aurorae, the Northern and Southern Lights.

Far-flung spacecraft, such as the two Voyagers and Pioneers 10 and 11, have found that the solar wind continues – albeit more slowly – right through the Solar System. They have picked up echoes from further out, which may be generated where the solar wind hits the gas between the stars. Within this boundary, the heliopause, space is filled with the Sun's outflowing gases: all the planets are effectively orbiting within the Sun's atmosphere.

The Sun's huge outer atmosphere of hot gases, the corona, is a million times fainter than the brighter photosphere. It is normally invisible, because it is only one hundredth as bright as the daytime sky. But an eclipse shows the corona in its true glory (**Fig. 2.13**) when the Moon blocks out the photosphere's light, and with it the blue sky of scattered sunlight. The corona's brightness falls off sharply with distance from the Sun, and this photograph compensates with a filter which is dark at the centre and progressively more transparent to the edges. As a result, details are visible from the inner atmosphere, the chromosphere, to a distance of two Sun diameters – 3 million kilometres. The chromosphere displays reddish prominences at top left; above, in the lower corona, are loops and 'helmets' of gas held

2.16 *Ultraviolet, 365 nm, Pioneer Venus Orbiter cloud photopolarimeter*

2.17 *Infrared, 2.4 μm (green) and 5 μm (red), Anglo-Australian Telescope*

2.18 *Infrared, 2.3 μm, Galileo Near Infrared Mapping Spectrometer*

Venus

WITH A DIAMETER almost as large as the Earth, Venus has long been regarded as our twin. Although it is the planet that can come closest to the Earth (within 40 million kilometres), until recent years it was also the most enigmatic. Optical telescopes show us nothing of Venus's surface, because the planet is completely covered in brilliant featureless clouds.

It has taken the full power of the new astronomies – spacecraft and fresh kinds of observation – to strip away the veils of Venus. To many, the result has come as a disappointment. Venus turns out to be the hottest, driest and most hostile of all the planets – the closest place to Hell we know in the Solar System.

Fig. 2.16 is a close-up view of the tantalising clouds of Venus in ultraviolet light, taken by the Pioneer Venus Orbiter which went into orbit around the planet in 1978. It shows the uppermost of Venus's three cloud layers, some 65 kilometres above the planet's surface. The swirling patterns appear because ultraviolet is absorbed by various sulphur compounds in the clouds, which are made mainly of concentrated sulphuric acid. By watching these patterns, astronomers have found the clouds move round the planet in four Earth-days, far more quickly than the planet's own rotation period of 243 days.

Using an Earth-based telescope, the British astronomer David Allen was surprised to discover, in 1983, infrared radiation coming from the dark side of Venus. Until then, astronomers had thought that the clouds were totally opaque to radiation trying to escape upwards from the hot lower regions of the atmosphere.

Fig. 2.17 is an infrared image of the dark side of Venus. (The crescent-shaped sunlit region lies to the left and has been cut off,

with the exception of the bright white cusp at the bottom.) The red regions show the long-wavelength emission from the cool upper cloud decks, while green reveals the radiation coming from deep layers where the temperature is 450 K (180°C). The patchiness is caused by structure in the middle layer of Venus's clouds.

This structure shows up in a more detailed infrared image (**Fig. 2.18**) obtained by the Galileo spaceprobe as it swept past Venus in 1990, on its way to Jupiter. The colour coding shows the brightness at just one infrared wavelength, from white for the brightest through red, yellow, green and blue to black for the faintest. These patterns reveal clouds about 50 kilometres above the surface, in Venus's main cloud deck. To north (top) and south (bottom) the clouds are drawn into filaments by winds blowing at 240 kilometres per hour.

2.19 *Radar, 12.6 cm, Magellan Synthetic Aperture Radar*

In **Fig. 2.19**, we are flying over the hot lava plains of Venus towards the giant volcano Maat Mons, which appears on the skyline 1000 kilometres away. It looks like a scene from a science fiction movie, but this is science fact: the combined power of radar and computers has shown us how Venus's surface really looks under the all-enveloping clouds.

Radio waves can easily penetrate Venus's clouds and its thick atmosphere. Since the 1960s, astronomers on Earth have been directing radio waves to Venus and trying to determine, from the faint returning 'echo', what kind of a world is reflecting these beams. One of the first results was to find that Venus rotates very slowly, and in the opposite direction from most of the other planets.

But detailed views had to await spacecraft that could take radar sets into orbit around Venus. The American Pioneer Venus Orbiter blazed the trail in 1978, followed by the Soviet Venera 15 and 16.

The breakthrough came with the American probe Magellan, which arrived at Venus in 1990. Its radar could reveal details only 100 metres across – as small as a city block.

With computers on Earth, scientists could turn Magellan's raw data into views like Fig. 2.19. The colour of the surface has been chosen to match views from Soviet craft that actually landed on Venus and sent back colour pictures. There is one element of artistic licence here: the exaggerated relief, which makes mountains appear 20 times higher than they really are.

Much of Venus is covered with the rolling lava plains seen in the foreground of Fig. 2.19. Impact craters like the specimen in the centre of this view – made by an infalling meteorite – are very rare. All but the largest meteorites are burnt up as they fall through the thick atmosphere, some 90 times denser than Earth's. But geologists believe that craters are also uncommon because fresh lava flows and volcanoes have covered many old craters. By counting the

impact craters – Magellan discovered 912 – geologists conclude that most of the present surface of Venus was laid down only 500 million years ago, less than one-tenth of the planet's age.

Some researchers believe that a great geological cataclysm resurfaced Venus at that time with new volcanoes, like Maat Mons, and floods of lava. Indeed, such episodes might have repeated every few hundred million years, as heat built up under Venus's thick crust. Other geologists believe the cratering record can as well be explained by a fairly continuous level of volcanic activity.

Either way, Venus has been an immensely active world. In the planet's early days, its volcanoes undoubtedly produced the carbon dioxide that now forms the planet's dense atmospheric cloak. The pressure at the surface is 90 times Earth's atmosphere pressure. This gas in turn creates a powerful greenhouse effect, trapping the Sun's heat and raising the

2.20 *Radar, 12.6 cm, Magellan Synthetic Aperture Radar*

surface temperature to 465°C – hot enough for the rocks to glow dull-red in the dark.

During one Venusian year – 225 Earth-days – Magellan mapped almost the entire surface of Venus. This information was put together to make **Fig. 2.20**, a view of one hemisphere of Venus as it would look if we could strip away the clouds and atmosphere entirely.

Stretching across the middle of this view, practically along Venus's equator, is a volcanic region called Aphrodite. Standing 3000 metres above the surrounding plains, Aphrodite is studded with volcanic cones

and surrounded by deep curving canyons. To the North and South are dark lava plains. (The black triangular region at bottom right was not mapped by the Magellan spacecraft.)

Even geologically, Venus turns out not to be the Earth's twin. If we strip away the Earth's oceans and clouds, there is a distinct pattern of old highland regions (continents), towering over the much younger plains that form the ocean floors. The Earth has an obvious worldwide pattern of volcanoes and ridges, marking the edges of huge 'plates' on which the continents gradually move around the globe.

There is only one small region on Venus that looks even remotely like a terrestrial continent – a plateau near the north pole called Ishtar. The crust of Venus is certainly not split into distinct plates, and there is no movement of the surface around the planet. Perhaps Venus's hot crust is not rigid enough to shift around in large plates, and just crumples up locally when volcanic forces operate. Or perhaps the volcanic forces on Venus recycle its crust by vertical movements instead of horizontal motions. Even after Magellan has stripped away its veils, Venus has yet more secrets to reveal.

2.21 *Optical, true colour, Voyager 1*

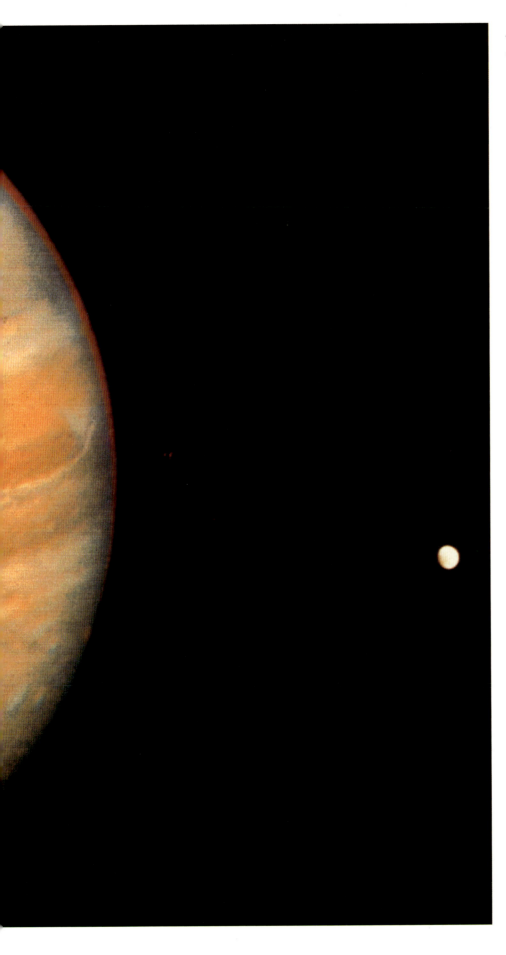

Jupiter

THE GIANT OF the planets, Jupiter, has a diameter of 11 Earth-diameters, and a mass 318 times that of our planet. Jupiter may have a rocky core, similar in size and composition to the Earth, but most of its bulk is composed of lighter substances, mainly hydrogen and helium, but with some methane, ammonia and water. The planet's gravity compresses its own matter so that it behaves like a liquid rather than a gas, apart from a thin outer gaseous atmosphere containing layers of coloured cloud.

Jupiter is large enough that its general shape and features are easily seen with Earth-based optical telescopes. The planet appears noticeably flattened, because its equatorial regions bulge outwards under the effect of the 'centrifugal force' due to Jupiter's fast rotation. Turning once in only 9 hours 55 minutes, Jupiter rotates faster than any other planet. The main features seen on Jupiter are bands of different coloured clouds running parallel to the equator. Within these bands are finer streaks and ovals. The largest oval, the Great Red Spot, is very long-lived for an atmospheric feature, having been observed as early as the 1660s. Earth-based telescopes have also revealed a dozen satellites; four are comparable in size with the planets Pluto and Mercury.

But our knowledge of Jupiter has recently been revolutionised by spaceprobes studying the planet and its satellites at close quarters. **Fig. 2.21** is one of the 35 000 photographs returned by the two Voyager probes in 1979. At the extreme right is the white satellite Europa, a smooth ice-covered world; while in front of Jupiter is the orange moon Io, coated in sulphur ejected by its active volcanoes – one of the major discoveries of Voyager 1.

Sequences of Voyager photographs reveal that the light and dark bands are travelling around the planet at different speeds, up to 400 kilometres per hour, with bands at successively higher latitudes moving round in opposite directions. The ovals lie between these oppositely directed wind currents, and rotate with the relative motion – like ball bearings between two moving surfaces. The Great Red Spot (lower left in Fig. 2.21) seems to be just a particularly large oval.

Fig. 2.21 shows that the clouds also contain turbulent swirls, or eddies. These eddies are transporting energy upwards within the atmosphere, and meteorological studies have indicated that it is the energy of the eddies which is eventually harnessed to drive the regular wind currents at different latitudes. The Earth's atmosphere also has jet streams fed by the energy of turbulent eddies, and studies of Jupiter's atmosphere are providing clues to the Earth's weather and climate.

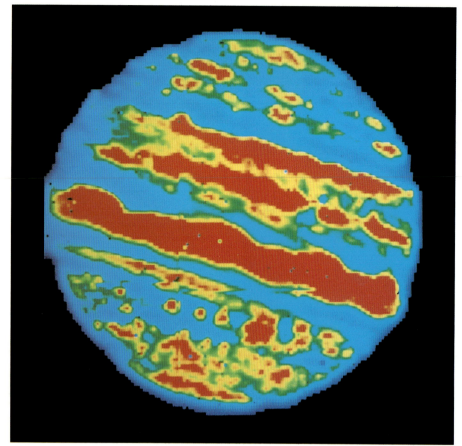

2.22 *Infrared, 4.8 μm, 3 m NASA Infrared Telescope Facility*

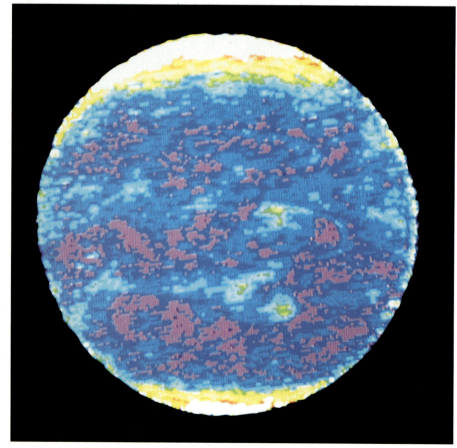

2.23 *Optical, 600 nm, linear polarisation, 1 m telescope, Meudon Observatory, France*

Observations of Jupiter at other wavelengths do not reveal the stunning details of the visual images from spaceprobes, but they show regions – both above and below the visible clouds – which we cannot investigate using ordinary light. Earth-based 'new astronomy' images are complementing what we have learnt from spacecraft.

In the infrared (**Fig 2.22**), Jupiter is almost a negative of its optical image. The white cloud zones now appear dim, and the brown bands to either side of the equator shine brightly at infrared wavelengths. The faintest regions in this infrared image are coloured blue, and brighter parts successively green, yellow and red.

The fainter parts of this image consist of sunlight reflected from the high white clouds, but the brighter portions show us heat radiation shining out from the interior of the giant planet. We detect this emission from regions of Jupiter where the deeper layers are not covered by the higher clouds.

The white clouds are made of frozen ammonia crystals, at a temperature of 130 K. In the gaps, we look down to a deck of brown clouds about 50 kilometres lower down. According to the infrared measurements, these are at a temperature of 230 K (no colder than an arctic winter). Given their temperature, the appropriate pressure and Jupiter's overall composition, it is most likely that these clouds are made of ammonium hydrosulphide, a basically white substance tinged brown with traces of other sulphur compounds.

To keep at a steady temperature, a planet should radiate away as much energy as it receives from the Sun. But the infrared emission from Jupiter amounts to 70 per cent more than this. The planet must therefore have its own interior source of heat. It is probably kept warm because it is still contracting slightly from its formation out of a gas cloud, and the consequent compression heats its interior.

Even deeper and hotter regions of Jupiter can be glimpsed through small holes in the deck of brown clouds. These regions, as warm as 'room temperature' (300 K), show as brilliant spots on infrared maps. They can also be recognised on Voyager images as hazy blue spots, because light from below is scattered by the denser gas, much as Earth's air scatters sunlight to cause a blue sky.

This scattering shows up more clearly when we study the polarisation of light reflected from Jupiter. Light is an electromagnetic wave, vibrating at right angles to its direction of travel. Looking at a beam of light end-on, the orientation of its vibration is the direction of polarisation. Sunlight is 'unpolarised' – it consists of a jumble of all possible orientations – as is sunlight reflected from a cloud. But sunlight penetrating deep into an atmosphere becomes polarised as some of the radiation is absorbed. You can see this on Earth by rotating the lens of polaroid sunglasses as you observe the sky at sunset.

Jupiter appears in polarised light in **Fig. 2.23**. The false-colouring shows regions of low polarisation in pink, increasing through blue, green, yellow and red to white. Looking down into the two brown zones near the equator, the reflected light is clearly polarised (blue and green) – as it is around the Great Red Spot (lower right of centre).

But we see to the greatest depth into Jupiter's atmosphere at its poles. Here the polarisation of the planet reaches its maximum, around 1 per cent. The 'polar clearings', where high ammonia clouds are absent, were suspected as long ago as 1929, but this image taken nearly 60 years later was the first to show them clearly.

High above the polar regions, even more excitement is being generated. The infrared image (**Fig. 2.24**) is colour coded so that the fainter parts are blue, with brighter regions yellow and red. The bright blob in the centre is sunlight reflected off Jupiter's moon Ganymede, which happened to be passing in front of the planet at the time. But Jupiter itself is largely dark, because its abundant methane absorbs the 3.4-micrometre radiation used to make this image.

At the poles, Fig 2.24 shows Jupiter alight with radiation from its aurorae – the jovian equivalent of the Northern and Southern Lights on the Earth. Here, high-speed particles from space are crashing into the top of Jupiter's atmosphere. Ordinary cameras on Earth cannot see Jupiter's aurorae, but this observation has tuned into the radiation emitted by a rare kind of molecule, a trio of hydrogen atoms with a positive electric charge. These unstable molecules are created in the aurorae, as the normal two-atom hydrogen molecules are broken up and recombine.

The particles that create the aurorae are channelled towards Jupiter's poles by the planet's strong magnetism – as is clearly revealed in an X-ray image of Jupiter (**Fig 2.25**). The planet's disc is marked by the yellow oval, and the plane of its equator by the straight lines. The X-ray emission is coded dark blue for the faintest regions, and pale blue and white for the more intense parts. The upper regions of the aurorae can be seen stretching well away from the planet, into its tenuous outer atmosphere. The aurorae lie above the planet's magnetic poles, which – as can be seen here – lie some 10° from its pole of rotation.

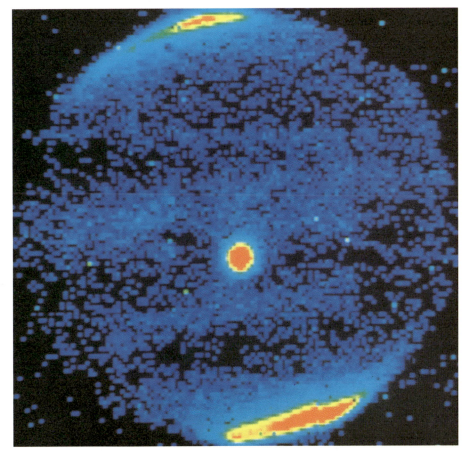

2.24 *Infrared, 3.4 μm, 3 m NASA Infrared Telescope Facility*

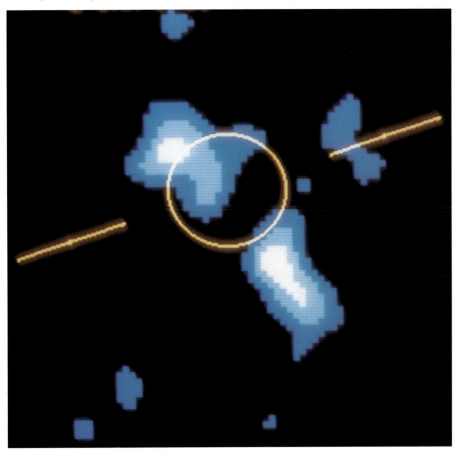

2.25 *X-ray, 0.4–8 nm, High Resolution Imager, Einstein Observatory*

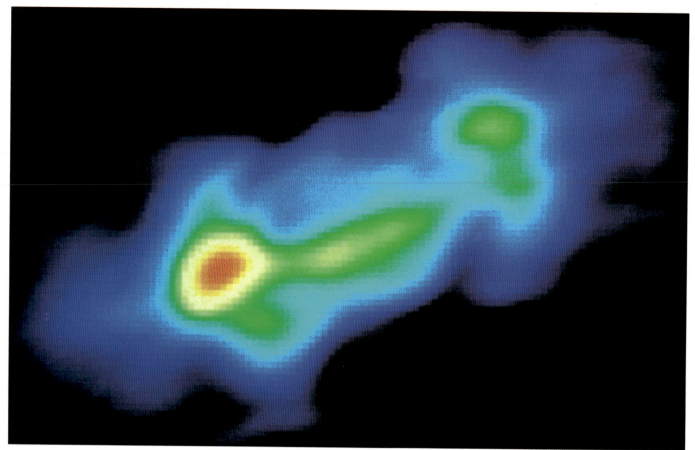

2.26 *Radio, 20 cm, Very Large Array*

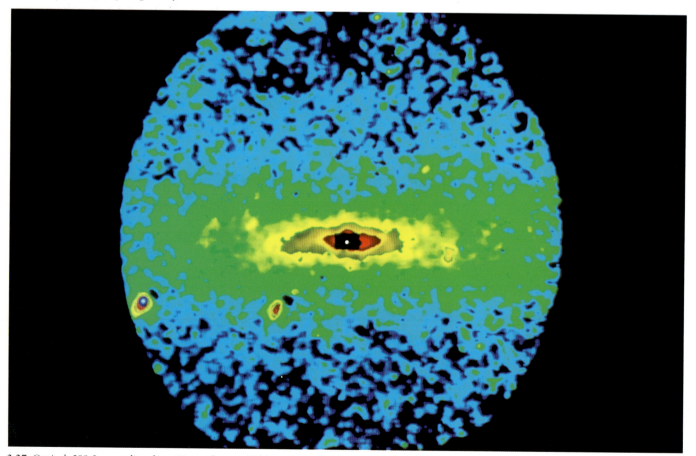

2.27 *Optical, 589.3 nm sodium line, 61 cm telescope, Table Mountain Observatory, California*

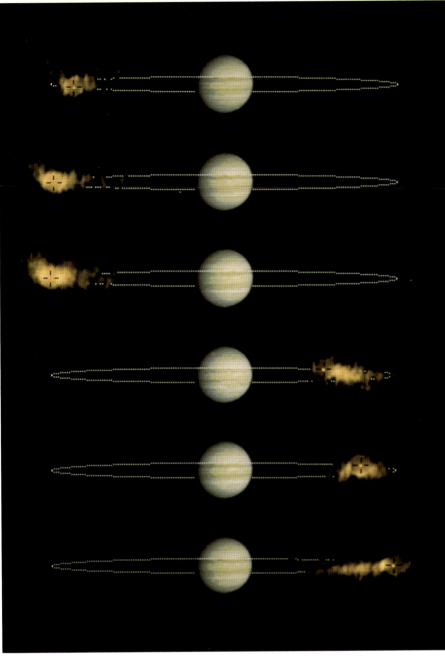

2.28 *Optical, 589.3 nm sodium line, 11 cm telescope, McDonald Observatory, Texas*

Jupiter is a brilliant source of radio waves when observed from the Earth. These emissions come from outside the planet itself, arising in a huge region of magnetism – the magnetosphere – that surrounds Jupiter.

At relatively short radio wavelengths (**Fig. 2.26**), Jupiter looks triple. The colour coding here indicates faint regions in dark blue, and brighter portions of the image in pale blue, green, yellow and red. The planet itself appears only faintly, as the central blue oval with a green stripe. This radiation is a result of Jupiter's innate warmth – an extension of its infrared emission to slightly longer wavelengths.

More prominent are the 'wings' to either side – especially the yellow- and red-coded wing to the left. These are caused by Jupiter's intense magnetic field. The field, about 20 000 times stronger than the Earth's magnetism, traps fast electrons in a belt around Jupiter which is similar to the Earth's Van Allen belts. As the electrons whirl through the magnetic field, they produce radio waves by the synchrotron process. As a result, Fig. 2.26 maps the inner part of Jupiter's magnetosphere.

The regions of the belt differ in brightness in Fig. 2.26 because the magnetic field concentrates the synchrotron radiation into narrow beams. At the left side of Jupiter, the magnetic field is beaming the strongest radiation directly towards the Earth, while the field on the right sends the beam in a direction that misses the Earth, so this region appears faint.

The first radio emissions to be detected from Jupiter, in 1955, were at much longer wavelengths, around ten metres. These erratic 'decametric' signals come largely from thunderstorms and aurorae in Jupiter's atmosphere, but they tend to repeat with the same period as Jupiter's moon Io revolves around the planet. It was the first hint that space round Jupiter is far from empty.

Io itself is responsible for much of this pollution. **Fig. 2.28** is a series of images taken through a filter that passes only the yellow light from sodium atoms; Jupiter and Io's orbit have been superimposed, with the position of Io marked with a cross. As Io moves around Jupiter, it is accompanied by a huge banana-shaped cloud of sodium atoms stretching both ways along its orbit.

This moon, it turns out, orbits Jupiter well within the planet's large magnetosphere. As a result, Io is bombarded by high-speed particles that chip atoms off its surface. Along with the occasional impact of meteorites and volcanic eruptions on Io, this constant erosion provides Io with a tenuous atmosphere made of potassium, oxygen, sulphur and sodium – which happens to be the easiest to detect. These gases can escape from Io's weak gravity, and spread out along its orbit.

More sensitive detectors show these sodium atoms spreading much further away. **Fig. 2.27** was again taken through a filter that passes only sodium light, but covering a region reaching to 70 times the size of Io's orbit. Here, colour coding shows the greatest concentration of sodium in red, decreasing through orange, yellow and green to blue for the lowest levels. Jupiter and Io themselves are both hidden by the central white spot.

The densest parts of the sodium cloud (red and orange in Fig. 2.27) show the size and shape of Jupiter's magnetosphere, stretching out a distance equal to 100 radii of the planet itself.

But the electrically-neutral atoms of sodium are not strongly controlled by magnetism, and they can easily leak out beyond the edge of the magnetosphere. The result is a giant 'magneto-nebula', out to the edge of the green region in Fig. 2.27, which extends to at least 400 Jupiter-radii. This makes it 50 times larger than the Sun, and by far the largest structure in the Solar System. If our eyes could tune in to this faint sodium emission, we would see a glowing yellow cloud surrounding Jupiter that is 6° across – over ten times the size of the Moon in the sky.

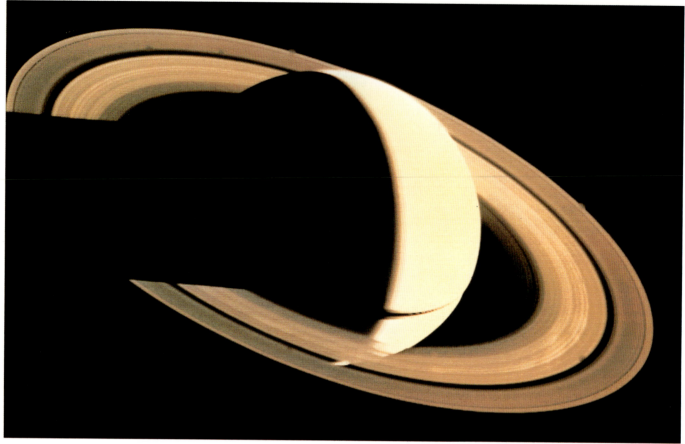

2.29 *Optical, true colour, Voyager 1*

Saturn

SECOND ONLY TO Jupiter in size and mass, Saturn is a giant planet almost ten times as wide as the Earth and 95 times as massive. Its fame, however, rests not on its dimensions, but on its rings. Although Jupiter, Uranus and Neptune have faint ring systems, Saturn's wide bright rings are unique, making it the most beautiful planet as seen by telescope or from the passing spaceprobes Voyagers 1 and 2.

Fig. 2.29 was taken by the Voyager 1 probe four days after it had passed closest to the planet. The spacecraft was able to see Saturn as we can never view it from the Earth, with the planet only half-illuminated by the Sun. Saturn's shadow is thrown across the rings behind, while the rings in front of the planet cast shadows onto Saturn's sunlit hemisphere.

The rings' structure is seen in Fig. 2.29 much as it appears from Earth – except in far finer detail. A dark band running round the planet divides the system into two main rings. Outside this dark Cassini Division lies the A-ring; while inside is the brighter B-ring. Within the B-ring, Fig. 2.29 shows the faint C-ring – or crepe ring – best seen just to the right of the centre of the planet's lit-up hemisphere.

Earth-bound observations had indicated that the rings are not solid objects girdling Saturn, but must consist of huge numbers of small ice blocks each orbiting the planet as a small moon. Even the Voyagers' close up pictures could not show the individual blocks, but they did reveal – to the amazement of the astronomers – that this ring material is concentrated in about 100 000 individual ringlets. Fig. 2.29 shows some of the more prominent ringlet structure in the B-ring – although each 'ringlet' seen here is in fact composed of hundreds of narrower ringlets. The picture also demonstrates dramatically that the rings are indeed not solid: at the bottom, Saturn's globe is visible through even the densest part of the B-ring.

The Voyagers could determine the size of the blocks comprising the rings in several ways. Their effect on the Voyagers' radio transmissions as they passed behind the rings indicate that the main blocks are about 1 metre across in the C-ring, and are progressively larger farther out from the planet, to reach 10 metres in the A-ring. They are separated by a distance roughly ten times the block's diameter.

The ice blocks appear yellowish-white in true colour pictures like Fig. 2.29, but when the colour differences are extremely enhanced by computer (**Fig. 2.30**), the rings are seen to differ slightly in colour. The C-ring (left) and the particles in the Cassini Division (towards top right) reflect short wavelengths better, and appear dark blue in Fig. 2.30. The B-ring changes colour smoothly from the outside (blue-green) to the inside (orange), which reflects long wavelengths most efficiently. There is a surprisingly sharp change in colour between the B- and C-rings, much more pronounced than the change in brightness (compare with Fig. 2.29).

Since the Voyagers flew past Saturn, Voyager 2 has also sent back images of rings surrounding Uranus and Neptune. These are narrow and very dark. The icy particles in these rings were probably once as brilliant as fresh snow, but they have been darkened by the impact of charged particles and micrometeorites.

The implication is that the bright rings of Saturn must be comparatively young. The brightness and colour of the B-ring indicates the particles here have only been exposed for 100 million years – a fraction of the planet's age. At the time when dinosaurs were roaming the Earth, an icy moon of Saturn about 100 kilometres across must have been struck by a comet. Its fragments spread out to form the planet's spectacular rings.

While the Voyagers produced by far the most detailed views of Saturn ever seen on Earth, they provided basically just two snapshots. Astronomers have had to rely on telescopes on – or near – the Earth to keep track of changes on the planet.

Most spectacular are the 'white spots'. Every 30 years or so – in 1876, 1903,

2.30 *Optical and ultraviolet, enhanced colour, Voyager 2*

1933, 1960 and 1990 – astronomers have seen a brilliant white cloud appear in the planet's generally yellow cloud-tops. The eruptions come when the planet is in the same point in its 30-year orbit, when the North Pole is most tilted towards the Sun, and may be a result of the extra solar heat on the atmosphere at the pole.

In 1990, the Hubble Space Telescope observed the white spot in unprecedented detail and at different wavelengths. The false-colour **Fig 2.31** is a combination of Hubble's view through a blue filter (coded blue) and through a far-red filter (coded red) at a wavelength absorbed by methane and so revealing only the high clouds above the planet's methane-rich atmosphere. Hence the lower regions of cloud appear blue and higher parts orange (the bright red region corresponds to the C-ring in front of the planet).

In this view, taken on 9 November 1990, six weeks after the outburst was first spotted, the 'white spot' has spread out all around the equator of Saturn. The clouds seen as brilliant white to the human eye appear orange in this colour coding, showing that they lie above the planet's ordinary cloud-decks. The northern edge has been sculpted into ripples by the planet's hurricane-force winds. Astronomers believe the eruption is a bubble of ammonia which burst through the lower clouds and froze into white crystals high in Saturn's atmosphere.

2.31 *Optical, 435 nm (blue), 815 nm (red), Wide Field/Planetary Camera, Hubble Space Telescope*

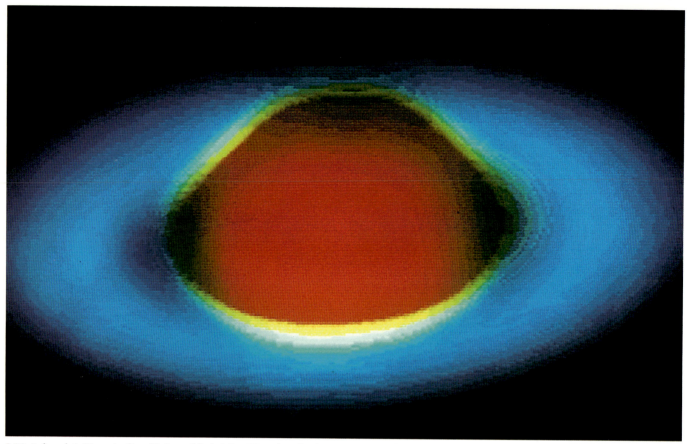

2.32 *Infrared, 1.65 µm (red) and 2.2 µm (blue), 3.8 m UK Infrared Telescope*

The difference between gaseous Saturn and its icy rings – both yellow-white to human eyes – shows up dramatically when the planet is seen in near-infrared radiation. **Fig. 2.32** is a composite of two views through filters that pass wavelengths near 2 micrometres, one of which is strongly absorbed by methane gas.

The rings reflect radiation of all wavelengths, and appear blue in this composite. But the gaseous planet absorbs sunlight with a wavelength of 2.2 micrometres, so it appears in Fig. 2.32 only in the reflected radiation at 1.65 micrometres, coded red.

At longer infrared wavelengths, we do not have to rely on reflected sunlight: we can detect Saturn by the emission of its own heat energy. **Fig 2.33** is a false-colour view of the planet's globe as seen at a middle infrared wavelength that is emitted by methane gas. The lower part of the globe is cut off by the rings running across the bottom of this view as a curved black frame to the bright image. Saturn's North Pole lies at the centre of the coloured ovals. The false colour shows the dimmest – and therefore coolest – regions in orange, and successively brighter and warmer regions in yellow, green, blue, pink, red and white.

In Fig. 2.33, a narrow warm band (blue and green) is visible around Saturn's equator, and an even warmer region (white, yellow and red) right at the North Pole. In these regions we are observing gas at a lower level than the cold cloud-tops seen in ordinary light.

The warmth of Saturn's interior shows up even more clearly when we move to radio wavelengths. In the radio image (**Fig. 2.34**), the cold rings appear dim (coded blue) while its natural heat radiation makes the globe bright (coded orange and red). We look deepest into the planet near the centre of its disc, so this region appears hottest (red). Towards the edge of the disc, our oblique line of sight penetrates only the cooler upper layers (orange). As a result, the radio Saturn is 'limb-darkened', like the Sun's appearance at optical wavelengths (Fig. 2.2) – though the colour coding here seems to show the opposite!

The supply of heat which makes Saturn glow in Fig. 2.34 is probably a result of rearrangements deep in its interior. Like Jupiter (Fig. 2.21), Saturn is made up largely of hydrogen and helium – substances with such a low density, even when compressed into a planet, that Saturn would float in water, given an ocean large enough! While Jupiter's interior heat comes from slow shrinkage, originating from the time when the planets first formed some 4600 million years ago, calculations show that the smaller Saturn should by now have finished its contraction and should have cooled down throughout.

Saturn's heat must come from a different process. At its formation, the hydrogen and helium atoms would have been well mixed together throughout the planet. But Saturn's core has now cooled to the point where the atoms would naturally separate, like oil and vinegar in a salad dressing. The less-abundant helium atoms should be gathering together as helium droplets within the surrounding hydrogen. Because helium is denser than hydrogen, these droplets fall towards Saturn's centre, to make a helium core within the predominantly hydrogen planet. In their fall, the droplets lose gravitational energy, which is converted to heat.

This theory would mean that Saturn's outer layers should by now have been drained of about half their original helium. The Voyager spaceprobes have indeed confirmed this prediction. While Jupiter's atmosphere contains 11 atoms of helium to every 100 hydrogen atoms – like the Sun and gas clouds in space – Saturn has only 6 helium atoms to 100 atoms of hydrogen. Hence it seems that Saturn's infrared glow is the result of helium rain in the planet's interior, dripping into a hidden central helium ocean larger than planet Earth.

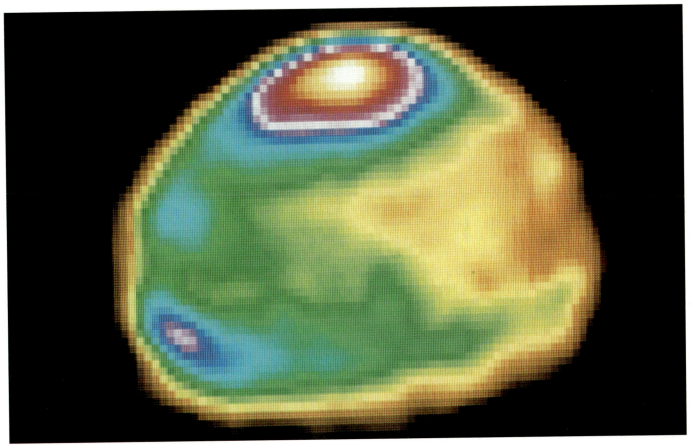

2.33 *Infrared, 7.8 μm, 3 m NASA Infrared Telescope Facility*

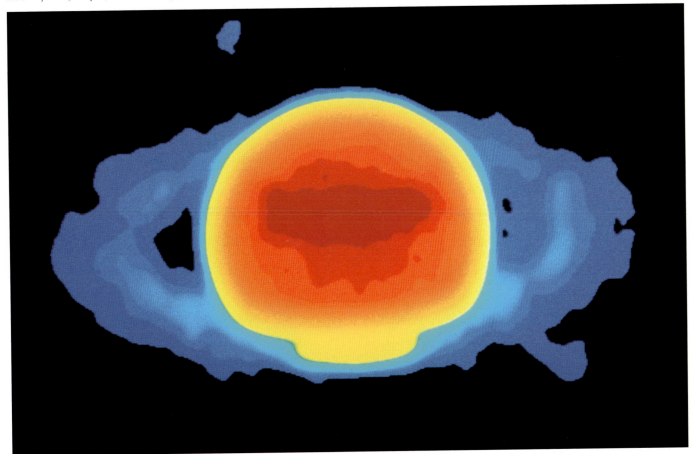

2.34 *Radio, 6 cm, Very Large Array*

2.35 *Optical (14 April 1986), true colour photograph with wide-angle lens, New Zealand*

2.36 *Ultraviolet (13 March 1986), 122 nm, rocket-borne telescope*

Halley's Comet

TRULY A 'ONCE-IN-A-LIFETIME' experience', Halley's Comet enlivens the skies of planet Earth once every 76 years. In between times, it spends decades near the far point of its long oval-shaped orbit, as far from the Sun as Neptune. Only when Halley swings in close to the Sun does it grow a gaseous head and long cometary tail.

Chinese astronomers recorded this comet as early as 1059 BC. In 1682, it was observed by the English astronomer Edmond Halley. He realised the same comet had been seen in 1531 and 1607, and correctly deduced that comets follow closed orbits around the Sun. This comet, Halley predicted, would return in 1758. When it appeared on schedule, the comet was naturally named after him.

Astronomers scrutinised Halley's Comet on its returns in 1835 and in 1910. But when it appeared in 1985–86, the comet was subjected to two new and very powerful techniques: telescopes that could observe radiations other than visible light,

and a fleet of visiting spacecraft.

Passing by the Earth in April 1986 (**Fig. 2.35**), Halley's Comet looked the archetypal comet, with its glowing head and straight tail – rather foreshortened in this head-on photograph.

Fig. 2.35 also provides a lesson in cosmic perspective. Halley's Comet is by far the nearest of three fuzzy-looking objects. At 'only' 66 million kilometres from the Earth, the comet's light took just under four minutes to reach the photographic emulsion. At the bottom of the shot is a giant cluster of stars, omega Centauri, whose light had been travelling for 16 000 years. And appearing next to Halley's Comet is a galaxy far beyond our own, Centaurus A (Fig 12.2), whose light had been on its way for 16 *million* years.

The blue colouration in the head and tail of Halley's Comet is caused by carbon monoxide gas fluorescing in the Sun's bright light. These molecules have been stripped of electrons, so they are positively charged, and the solar wind (Fig. 2.15) sweeps them up in a glowing tail that

always points away from the Sun. In Fig. 2.35, the comet is beginning its long trek back to the outer parts of the Solar System, travelling tail-first.

An ultraviolet view (**Fig. 2.36**) shows that the comet's head – or *coma* – is far larger than optical astronomers ever suspected. Here, the brightest region is in the centre, coded blue, with fainter regions successively yellow, green, black and red. These colours repeat twice more as we move outwards, with the faintest outer region also blue. This radiation comes from hydrogen atoms excited by the Sun. These atoms have no electric charge, and they are not affected by the solar wind. The coma in Fig. 2.36 is slightly elongated away from the Sun (towards the right) as the Sun's light exerts pressure on the atoms.

The ultraviolet coma seen in Fig. 2.36 would cover the entire area shown in the optical photograph (Fig. 2.35). Measured across the hydrogen coma, Halley's Comet is 25 million kilometres in diameter – almost twenty times larger than the Sun!

2.37 *Optical (13 March 1986), true colour, 15 cm telescope, Giotto*

On the night of 12–13 March 1986, the European spaceprobe Giotto shot right through Halley's Comet, providing the first close-up view of a comet's heart. As the only bright comet whose return can be predicted, Halley had for long been targeted for space missions. In the event, five spacecraft – from the Europe, Japan and the Soviet Union – were despatched in its direction.

Giotto came closest, passing just 605 kilometres from the nucleus. Its camera sent back images of a dark potato-shaped object, 8 kilometres wide and thick, and 16 kilometres long (**Fig. 2.37**). What was most surprising was its blackness: reflecting less than 3 per cent of the light falling on it, the nucleus of Halley's Comet is the darkest object we know in the Solar System.

As sunlight heats the comet's dark outer skin, the icy layers beneath begin to boil. As a result, brilliant jets of steam erupt on the sunward side of the comet. **Fig. 2.38** is a colour-coded version of one of Giotto's close-up views, showing details in the brightest jets. Here, the most brilliant part, in the centre, is coded white, with successively fainter regions in pale shades of orange, blue, pink and green, followed by

darker shades of pink, orange, yellow, red, pink, blue, green, orange, red, green, blue, green, orange, brown, purple and dark blue.

Part of the nucleus is visible as the dull dark region to the lower right. The prominent – and strongly-coloured – central region is a jet of steam, shining as the Sun's light catches it. The steam contains a small admixture of other gases, forming a noxious cocktail of formaldehyde, carbon monoxide and carbon dioxide, with a dash of ammonia, methane and hydrogen cyanide.

Once these molecules are out in the comet's coma, ultraviolet radiation from the Sun breaks them down into smaller fragments. These radicals are generally better at emitting radiation that can be detected by Earth-based telescopes.

Fig. 2.39 shows the jets of gas in the coma, as detected by a radio telescope tuned to radiation emitted by the hydroxyl radical (OH), which is a water molecule that has had one of its hydrogen atoms knocked off. Here, the brightest region is coded red, with fainter parts yellow, green and blue. The frame is centred on the position of the comet's nucleus,

and stretches 190 000 kilometres to either side.

The strongest emission is not in the centre in Fig. 2.39: it is displaced in the direction of the Sun, to the lower right. Here the erupting jets of steam have travelled some 50 000 kilometres from the nucleus and the water molecules are being split into hydrogen and hydroxyl. These radio observations show that Halley's Comet was shedding over 10 tonnes of water every second as it swung past the Sun.

The fainter blobs in Fig. 2.39 are the remains of jets that have erupted earlier, swept round by the rotation of the nucleus. The rate at which the comet's nucleus spins has been a matter of heated debate: the latest analysis suggests the nucleus is spinning round its long axis in 7.1 days and simultaneously tumbling every 3.7 days.

The spiralling path of outflowing gas is clearly seen in **Fig. 2.40**, a view of Halley in the blue light emitted by the cyanide radical (CN). The brightest parts are coded red and yellow, with fainter regions green, blue and shades of grey. The nucleus is at the centre, and the circular field stretches to a radius of 60 000 kilometres.

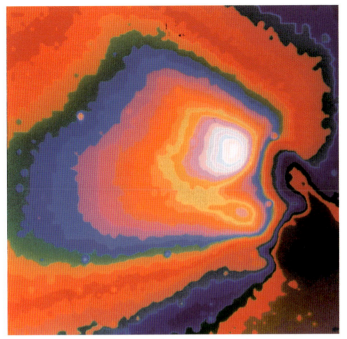

2.38 *Optical (13 March 1986), false colour, 15 cm telescope, Giotto*

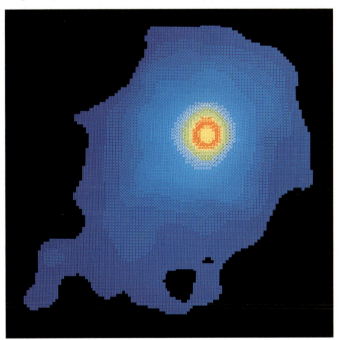

2.39 *Radio (13 and 16 November 1985), 18 cm hydroxyl line, Very Large Array*

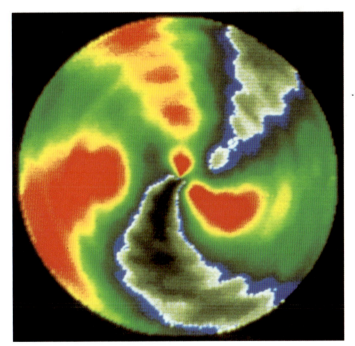

2.40 *Optical (20 April 1986), 387 nm CN line, 60 cm reflector, Perth, Australia*

2.41 *Infrared (18 November 1985), 10.8 μm, 3 m NASA Infrared Telescope Facility*

The jets of cyanide seen in Fig. 2.40 rather surprisingly do not coincide with the main jets of steam erupted from the nucleus. And they stay quite narrow, while the gas should spread out sideways to form broader clouds.

One explanation is that the cyanide jets are tracking the paths of very small dust particles. These 'CHON' particles – consisting of carbon, hydrogen, oxygen and nitrogen – were detected and analysed by Giotto and the Soviet Vega spacecraft. They could be erupting unseen from different vents on the nucleus. These solid particles keep to a narrow jet as they travel outwards. The Sun's ultraviolet radiation gradually breaks the CHON compounds down into radicals like CN, which show up the track of the parent dust grains.

Halley also ejects much larger dust grains. As these absorb energy from the Sun, they warm up and radiate the energy away again in the form of infrared radiation. **Fig. 2.41** is the first infrared image made of Halley's Comet. The false colours show the brightest region – coinciding with the comet's nucleus – in yellow, with red, green and blue depicting successively fainter regions further out. The distribution of dust in the coma extends to a diameter of 41 000 kilometres, with a trace of a tail to the lower left. The two Vegas and Giotto all suffered from thousands of impacts as they zoomed through the dust cloud at a relative speed of 240 000 kilometres per hour. One large grain tipped Giotto so that its antenna pointed away from the Earth for a crucial few minutes at closest approach, and damaged half of its instruments – where 'large' here means a mass of a thousandth of a gram!

3 optical astronomy

OPTICAL ASTRONOMERS HAVE built their biggest and best new telescopes far from civilisation and high above sea-level. The view of the skies from mountain peaks is unrivalled. Here above the turbulence, mists and clouds of the lower atmosphere, they can look up to a clear, black sky. The stars are steady, hard points of light, relatively free from the incessant twinkling familiar at sea-level as starlight is capriciously deflected by air currents in the lower atmosphere.

Moreover, the lower atmosphere is contaminated with background light – radiation from artificial lights scattered by minute dust particles in the air. 'Light pollution' makes the city nights so bright that the stars are swamped, and even in the countryside there is a slight background of

scattered light. It may be too faint to see, but when astronomers make long exposures to record the faintest possible stars and galaxies, the background light eventually builds up to fog the images and obscure the faintest objects.

So modern observatories are no longer built conveniently close to home. Instead, astronomers seek the best sites for astronomy. In the past, the world's major observatories were generally built where they were easily accessible. For example, King Charles II of England founded the Royal Observatory at Greenwich to assist navigators, and he placed it on a knoll overlooking the busy shipyards of Greenwich and Deptford on the River Thames, only a few miles from the royal court in London.

These days, however, astronomers are prepared to fly

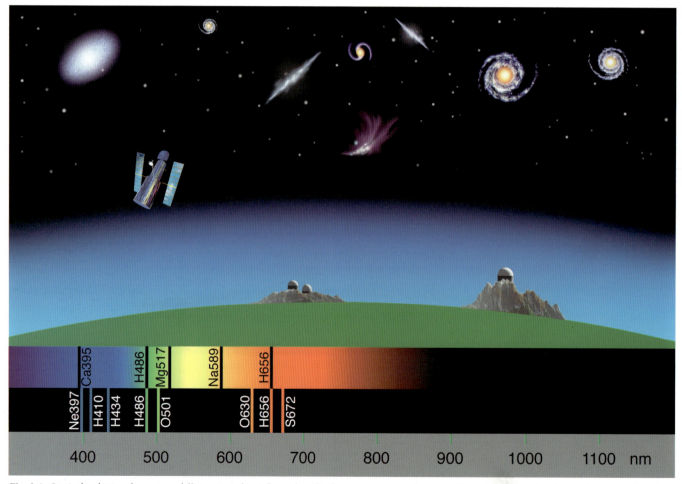

Fig. 3.1 *Optical radiation from space falls in a 'window' of wavelengths that can penetrate the atmosphere down to sea level, covering a wavelength range from 310 to 1100 nm (nanometre, equal to a millionth of a millimetre). The human eye can see only a narrower range of wavelengths (390 to 700 nm), the visible spectrum running through the rainbow colours from violet to red. The tenuous gas in a nebula produces bright emission lines at wavelengths unique to the elements in the gas. The most prominent lines come from neon (Ne), hydrogen (H), oxygen (O) and sulphur (S). The light from the dense gas in a star, in contrast, forms a continuous spectrum of all wavelengths, crossed by narrow dark lines where light is absorbed by elements such as calcium (Ca), hydrogen, magnesium (Mg) and sodium (Na). Optical telescopes – such as the Anglo-Australian Telescope and UK Schmidt (left) and the Keck Telescope (right) – are usually perched on mountains to get above the worst blurring from the Earth's atmosphere. From orbit, the Hubble Space Telescope (upper left) has the sharpest view of stars, nebulae and galaxies throughout the Universe.*

Fig. 3.2 *The 4200-metre volcanic summit of Mauna Kea, on the island of Hawaii, boasts the most powerful assembly of telescopes in the world. From left to right, the seven largest domes enclose the Canada-France-Hawaii Telescope (optical), University of Hawaii 2.2 m telescope (optical), UK Infrared Telescope, NASA Infrared Telescope Facility, Caltech Submillimetre-wave Observatory, Keck Telescope (optical and infrared) and James Clerk Maxwell Telescope (millimetre wave). Mauna Kea is dormant, but its neighbour Mauna Loa (on horizon) still occasionally erupts.*

halfway round the world, and take a jeep up remote barren mountains to reach their telescopes, and they have developed remote-control systems that allow them to operate a telescope halfway round the world from the comfort of their own office.

Probably the best observatory site in the world is the huge extinct volcano Mauna Kea (Fig. 3.2), which towers over the island of Hawaii. At its flat summit a huddle of telescope domes is growing amid a bleak lunar-type landscape of fractured lava, with no living plant or animal in sight. The University of Hawaii blazed the trail to this mountain peak; it now shares a large telescope here with Canada and France. The cluster of a dozen domes now includes the world's largest optical telescope: the mighty Keck Telescope with a mirror almost half the size of a tennis court.

The peak of Mauna Kea stands 4200 metres (14 000 feet) above sea-level. To take advantage of its unsurpassed view of the sky, astronomers have to face the problems of working at an altitude almost half the height of Mount Everest. The beaches of Hawaii may be a tropical paradise,

but the peak of Mauna Kea is always cold. In winter the observatory is buffeted by blizzards with winds of over 100 kilometres per hour; the observatory operates Hawaii's only snowplough! The air is so dry that skin and lips are constantly chapped – and the extreme dryness can play such havoc with electronic equipment that astronomers must sometimes boil kettles beneath the equipment racks to keep the air moist. Worst of all, though, is altitude sickness. The first dome on Mauna Kea was provided with oxygen equipment, but it proved too cumbersome to use, and not actually essential. Most astronomers find they can work in the thin, oxygen-poor atmosphere – but only slowly. Thinking becomes a problem. It can take a quarter of an hour to correct an equipment fault whose solution might be obvious in a minute at sea-level.

However, the rewards of working on remote inhospitable mountain peaks far outweigh the disadvantages. There the huge sophisticated modern telescopes seek out the faint rays of light coming in from planets, stars and distant galaxies.

Telescopes can gather light in two different ways

Fig. 3.3 *Different types of telescope all function to focus light onto a detector. A refractor (top left) uses a lens. Modern large telescopes, however, are reflectors (lower left), focusing light with a curved mirror. The light can be detected at the prime focus, in a cage supported within the top of the telescope's framework tube (inset). Alternatively, a secondary mirror placed just below this focus can reflect the light back through a hole in the centre of the main mirror, to the Cassegrain focus. A Schmidt telescope (top right) can 'see' a much wider region of sky. Its main mirror is part of a sphere (rather than a paraboloid) and a thin lens at the top corrects the distortions such a mirror would normally produce.*

Fig. 3.4 *The great '100-inch' Hooker Telescope on Mount Wilson in California was the world's largest from 1917 to 1948. Using a mirror instead of a lens to catch cosmic light, this reflector could be built far bigger than the refractors of the nineteenth century. The light grasp of its 2.5 m mirror enabled Edwin Hubble to make two major scientific discoveries: the Milky Way is but one of many similar galaxies; and the whole Universe is expanding, carrying distant galaxies away from us.*

(Fig. 3.3). A *refracting telescope* is the type that springs to most people's minds. Like the seaman's telescope of Nelson's day, a refractor has a lens at the top end of a tube, which collects light and focuses it to form an image at the lower end. The earliest telescopes were refractors. The most famous Renaissance scientist, Galileo Galilei, was the first to investigate the sky systematically with his little refractor – or 'optick tube' – which had a lens only 4 centimetres across. With this telescope he discovered the satellites of Jupiter and the phases of Venus, and resolved the Milky Way into stars.

Later refractors used larger and larger lenses, culminating in the huge Yerkes refractor near Chicago which has a lens slightly over a metre in diameter (40 inches). But there is a natural limit to the size to which you can build a refractor. The lens can be supported only at its edge, and a very large lens will sag out of shape under its own weight. The Yerkes lens weighs almost a quarter of a tonne, and is at just about the limit.

There is, however, another way to make a telescope, using a curved mirror (like a shaving mirror) to focus light to a focal point in front of it. The *reflecting telescope* has its mirror at the bottom of the tube. The tube – which may be just an open lattice work – supports some kind of device to intercept the focused light. In the very biggest telescopes, there is enough space for a small cabin at the focal point inside the telescope, where an astronomer can sit – making observations at the *prime focus*. More often, though, the tube is used to support a *secondary mirror* just below the

focus. This mirror reflects the focused image to a point where it can be examined more easily. In the small 'backyard telescopes' used by many amateur astronomers, the secondary mirror is set at an angle so that the image is formed at the side of the tube, and the astronomer looks into the side of the tube rather than through the end. This type is called a Newtonian reflector; it is the arrangement used by Sir Isaac Newton in the first reflecting telescope, which he produced in 1672.

The Newtonian arrangement is not very convenient for large telescopes, however, because the detector must be placed in an inconvenient and unbalanced position. The most common type is the *Cassegrain reflector*, in which a curved secondary mirror reflects the image back down the telescope tube and through a hole in the centre of the main mirror. The astronomer – or usually just a detector – is located behind the main mirror, looking up the telescope tube to the secondary mirror at a reflection of the sky reflected from the main mirror. (This design is now becoming popular with amateur astronomers too, and often there is a correcting lens at the top of the tube to make a *catadioptric telescope*.) Another alternative for large professional telescopes is to use a succession of mirrors to reflect the image down to spectrographs (see below) situated in rooms below the telescope dome.

Reflectors have one crucial advantage over refractors. A mirror can be supported over the whole of the back surface, and so it does not sag under its own weight. Sir William Herschel, the famous Hanoverian astronomer working in

Fig 3.5 *Ensconced within an art deco dome, the 5 m Hale Telescope on Palomar Mountain in California (unofficially known as the '200-inch') was the world's most powerful for several decades after its opening in 1948. The huge horseshoe at the right is part of a tilted mounting that allows the telescope to swivel round a single axis as the Earth rotates. This apparent X-ray view is a time exposure taken as the dome's narrow slit rotated. The Hale Telescope pinned down the first radio galaxies and quasars, and it has been a test bed for pioneering new light detectors, including the CCDs used on the Hubble Space Telescope.*

England, built a reflecting telescope as early as 1789 which was larger than the Yerkes refractor of a century later.

The first of the great modern reflectors (Fig. 3.4) was the 2.5-metre (100-inch) diameter telescope at Mount Wilson, California, designed by the great American astronomer George Ellery Hale and completed in 1917. Hale was convinced that an even larger reflector could be built, and he applied his persuasive powers to the Rockefeller Foundation. In 1948, ten years after Hale's death, the project was complete. A 5-metre (200-inch) telescope, named the Hale Telescope, was opened on Palomar Mountain in California (Fig. 3.5). In 1976, the Soviet Union completed a 6-metre telescope at the Zelenchukskaya Astrophysical Observatory in the Caucasus; but although it surpasses the Hale Telescope in sheer size the Russian giant is effectively less powerful because it lacks the electronics and computers to study its images in depth.

For 40 years after the completion of the Hale Telescope, however, most new telescopes were only about 4 metres in size. They included the 3.8-metre Mayall Telescope at Kitt Peak in Arizona, the 3.9-metre Anglo-Australian Telescope in New South Wales (Fig. 3.8), a 3-metre reflector at Lick Observatory in California, the 4-metre Cerro Tololo Interamerican Observatory reflector in Chile and the 4.2-metre William Herschel Telescope at La Palma. This trend away from larger telescopes came about because of new ways of detecting the light that reaches telescopes.

For the past century astronomers have only rarely looked through their telescopes. The human eye is simply not a very good light detector. It is sheer waste to put an astronomer's eye at the business end of an expensive, sophisticated telescope. The eye's main drawbacks are that it is not very sensitive to faint objects, and it cannot store the images it sees for analysis later; also the eye cannot measure brightnesses and positions of stars with any precision. For all serious astronomy, telescopes now pour their light onto an electronic detector – or, occasionally, a photographic plate. A modern telescope is less like Nelson's spyglass than a huge telephoto mirror system for a sensitive television camera.

For most of the twentieth century, photographic plates have been the astronomers' mainstay. They can expose a plate one night, and then take as long as they like to analyse the positions and brightnesses of stars, the shapes of galaxies, and so on. If they discover an interesting star or quasar, they can search back through old 'archival' plates to discover how bright it has been in the past, when it may have been recorded accidentally on a plate taken for another purpose.

There is a special type of telescope designed specifically to photograph the sky. The Estonian optician Bernhard Schmidt invented the type of wide-angle telescope now known as the *Schmidt telescope* (or Schmidt camera) in 1930. It is a strange cross between a reflector and a refractor (Fig. 3.3): a mirror at the bottom reflects starlight onto a curved photographic plate halfway up the tube, as in an ordinary reflector. But normally the images of objects at the edge of such a wide-angle view would be severely distorted. Schmidt corrected this distortion by putting a thin, slightly curved lens at the top of the telescope tube.

The first large Schmidt telescope – with a top lens 1.2 metres across – was installed at Palomar Mountain in 1948, to 'scout' for the Hale Telescope. Astronomers could search for interesting objects on the Schmidt plates, and then point the tunnel-visioned Hale giant to these particular objects. In the 1950s, the Palomar Schmidt was used to photograph systematically the whole of the sky visible from California. Each of its long-exposure plates recorded stars and galaxies as faint as magnitude 21 – a million times fainter than the dimmest star the eye can see. Each plate is 6° square – about twelve times the apparent width of the Moon. This Palomar Sky Survey gave for the first time a really detailed atlas of the sky to guide the new generation of giant telescopes. Similar Schmidt telescopes stationed in the southern hemisphere have now surveyed the regions of the sky which cannot be seen from California.

But there is a problem in reproducing astronomical photographs – whether from a Schmidt or from a conventional reflector. Modern plates can record outer faint tendrils of nebulosity which are ten thousand times dimmer than the nebula's bright central regions, and most of this brightness range is lost in printing. Either the faintest tendrils do not register, or the bright regions are overexposed.

One elegant solution is the technique of *unsharp masking*. The photographer uses the original plate – a negative – to produce a positive image on film. By putting this positive film mask over the original negative plate, the photographer effectively darkens the bright areas of the original negative image, and so decreases the range in contrast such that the faintest regions are only thirty times fainter than the centre of the nebula. In practice, astronomers like David Malin of the Anglo-Australian Observatory – a pioneer of the method – make the positive mask a little blurred ('unsharp'). This brings out the finest details in the original plate in crisp relief.

Faint details in photographs can also be revealed with the aid of computers. A laser beam scans back and forth across the photographic plate, in a two-dimensional 'raster' pattern like the lines on a television screen. A light-detector picks up the laser light that has come through the plate, converting it to an electrical signal that corresponds to the darkness of the emulsion at each point. This signal is analysed by a computer.

Laser-scanning devices like the Automatic Plate Measuring machine in Cambridge can scan through a whole Schmidt plate on its own, measuring the position and brightness of each of the millions of images. It can sort them into stars and galaxies, and determine the size and

Fig. 3.6 *An astronomer guides the 1.2-metre UK Schmidt Telescope at Siding Spring in New South Wales, Australia. This instrument has made a thorough photographic survey of the southern sky.*

shape of a galaxy. It then records all this information without the astronomer having to intervene in any way.

But photographic plates are now rarely used in astronomy, except for specialised tasks like sky surveys. In the 1980s, they were swept away by the electronics revolution. In place of silver salts developed by messy chemicals came a small robust silicon chip: the *charge-coupled device*, or CCD (Fig. 3.7).

Researchers started using electronic light detectors soon after the Second World War. The first were *photomultipliers*, simple devices which could measure the brightness of a star very accurately, but could do little more. They were succeeded by various types of electronic *image-tube*, which are highly sensitive television cameras. Light falls onto a sensitive metal surface in an evacuated glass tube, and it ejects electrons from the surface at a rate which depends on the light intensity at each point.

The CCD is more sensitive, and much smaller – roughly the size of a postage stamp. Researchers at Bell Telephone Laboratories invented the CCD as a miniature camera to be used in video-phones, but these never caught the public's imagination. CCDs have, however, become the heart of the home video camera.

Most silicon chips – used to control everything from washing machines to home computers – consist of a myriad of microscopic transistors and resistors in complex networks. The CCD is much simpler. It is an array of small light-sensitive regions called *pixels*. The image from a large modern telescope is focused onto the silicon slab, and

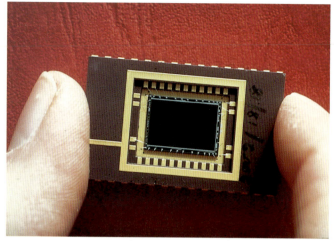

Fig. 3.7 *Spearheading the optical revolution is the CCD chip, the tiny 'retina' of today's huge telescopes. This example has 221 760 light-sensitive pixels in a matrix 385 down by 576 across.*

builds up a charge in each light-sensitive pixel proportional to the brightness of light falling on it. When the 'exposure' is complete, the charges can be read off into a computer memory, so that astronomers can analyse the image later.

The latest CCDs have over four million pixels, arranged in 2048 columns each containing 2048 pixels. They can register a view of the sky in far finer detail than a standard British television set (with 625 lines). They are ten times more sensitive than astronomical photographic plates in the blue to yellow parts of the spectrum, and compare even better in the red.

A home video camera produces a colour picture because it contains three CCDs, each seeing the view through a different coloured filter. For convenience and cheapness, astronomers use only a single CCD at a time, showing effectively a black and white view. When the astronomer calls up this image on a computer monitor, however, he or she will usually add false colours. As well as producing a gaudy image, false-colouring can bring out a wealth of scientific information.

The computer 'remembers' an image by storing a record of its brightness at each point. Instead of reproducing this information as a shade of grey on the monitor screen, colours can be assigned to each level of intensity: if the range of intensity were 100 to 1, the operator could assign red to the faint end of the range, say 1 to 25, yellow to the levels 26 to 50, green to 51 to 75 and blue to the brightest levels 76 to 100. The image on the screen is now multi-hued: all regions are the same brightness, but they differ in colour. The black of the background sky glows red; the most brilliant stars are blue – no brighter on the screen, but distinguished by a different hue – while regions of intermediate brightness form concentric green and yellow rings around the islands of blue.

This particular display would not tell the astronomer anything new. But in a search for very faint scraps of nebulosity, the four colours could be assigned differently, say to the levels 1, 2, 3 and 4, with the levels 5 to 100 black. Now all the bright images disappear from the screen as regions of uniform black: the background sky is still red, and the faintest wisps of nebulosity stand out in strident colour.

There are many variants on this theme, but the technique itself – *false-colour image-processing* – has become an essential tool for astronomers. It is also becoming more familiar on our television screens as an ingredient of pop videos. Image-processing is part of an increasing interaction between the astronomer and the raw data, with a computer as 'marriage-broker'. Instead of straining their eyes peering at a photographic plate, modern astronomers can sit in front of a colour monitor screen, and manipulate the image electronically. If they want to know the brightness of a star, they can point to it on the screen and the computer will work it out.

When observing the sky with a CCD (or a black-and-white photographic plate), astronomers usually put a colour filter in front, so they can be sure of exactly what colours (wavelengths) they are studying. How do you decide, for example, whether a particular bluish star is brighter or fainter than a neighbouring red star? Different light-detectors have different colour-responses from one another (and from the human eye), and so give conflicting results. The only answer is to observe through a filter which passes only one colour; and when comparing brightnesses, to stipulate which filter has been used.

The differences can be quite remarkable. The well known constellation Orion (the hunter) boasts two of the dozen or

Fig. 3.8 *The Anglo-Australian Telescope in New South Wales has a main mirror 3.9 metres in diameter, and a large 'horseshoe' equatorial mounting that turns gradually to compensate for the Earth's rotation. When opened in 1975, it was the world's most sophisticated telescope.*

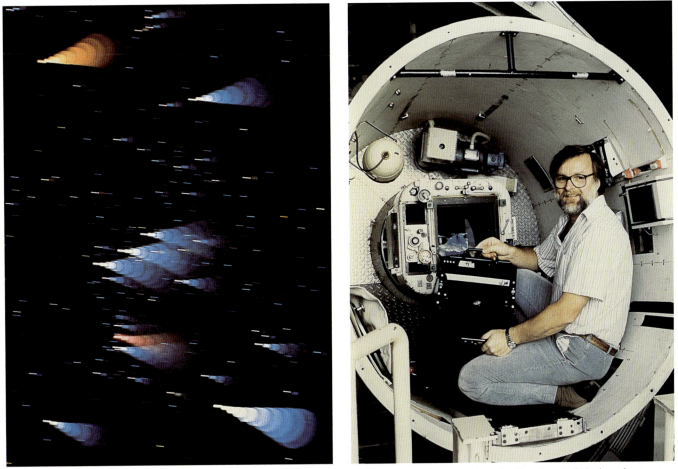

Fig. 3.9 *Colours of the stars show up more clearly when the light is trailed or defocused into a larger patch on the film – but the light from fainter stars may fade from visibility in the process. Here, David Malin has progressively defocused the constellation of Orion to show to advantage the colour from all its stars. Most are blue-white, indicating a temperature of 20 000 K or more. The brightest exception is the cool red giant Betelgeuse (top left), at only 3600 K. Below the three stars of Orion's belt is the Orion Nebula, shining pink in the light from hydrogen emitted at a wavelength of 656 nm.*
Fig. 3.10 *Sitting in the prime focus cage at the top of the Anglo-Australian Telescope is the world's leading astrophotographer, David Malin. He holds one of the large photographic plates that are exposed here, where light is focused directly by the telescope's main mirror.*

so brightest stars in the sky, reddish Betelgeuse at the top left of the hunter's outline, and bluish-white Rigel at the lower right (Fig. 3.9). But when seen (or photographed) through a blue filter, Betelgeuse becomes one of the fainter stars in the constellation, its intense ruddy glow extinguished by the filter. Conversely, a red filter shows Betelgeuse dominating the constellation; it dims Rigel's blue fire until it is no rival. If our eyes had evolved to be more sensitive to red than blue (or vice versa), we would have a very different idea of the hierarchy of the brightest stars in the sky.

The colours we perceive are a crude representation of the wavelength of the light radiation (Fig. 3.1). Our eyes are sensitive to a range of wavelengths from 390 nanometres (blue-violet) to 700 nanometres (red), a total range of some 310 nanometres. Photographic emulsions are also sensitive to somewhat shorter wavelengths, while CCDs are sensitive

to light slightly beyond the red end of the spectrum as seen by the human eye. Astronomers regard 'optical astronomy' as extending from 310 to 1000 nanometres (1 micrometre). The region that shows up on CCDs but is invisible to us – 700 to 1000 nanometres – is called the *far red*.

The filters normally used in astronomy pass quite a wide range in wavelengths: the standard 'B' (blue) filter, for example, lets through all radiation with wavelengths between 385 and 485 nanometres. The 'V' (for 'visual') filter passes yellow-green light, while the 'R' filter is transparent only to red light.

Just as a television picture is built up from three images shot in different colours, so astronomers can build up a true-colour picture of the sky from CCD images taken through three differently-coloured filters. These show the Universe to be full of colour. The hottest stars burn a fierce

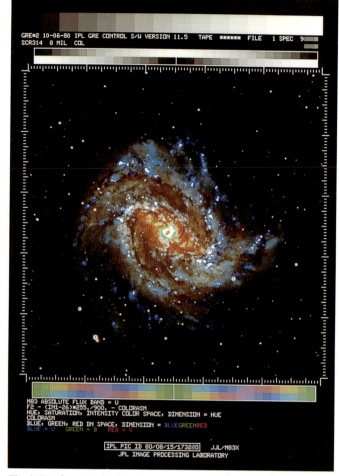

Fig. 3.11 *The nearby spiral galaxy M83, photographed with the 3.8-metre Mayall Telescope on Kitt Peak, Arizona. The image has been slightly computer-enhanced to reveal details (such as concentrations of hot young stars), but the colours are natural.*

Fig. 3.12 *Photograph of M83, image-processed so that views at different wavelengths appear in false colour. Red zones are regions of old stars and dust; green shows up ordinary stars; while blue areas are ultraviolet-emitting regions of very young stars.*

steely blue-white, while old cool stars glow orange or dull red. Nebulae come in a range of colours. Where their light comes mainly from hydrogen atoms, they shine a vivid crimson; where there is oxygen, the tint becomes apple green. Clouds of dust lit up by a star appear as sky-blue 'wraiths'.

Even to make colour pictures of the sky by conventional photography, astronomers take three black-and-white exposures through coloured filters, and combine them in the darkroom by printing each in turn through filters onto the same sheet of colour photographic paper. This rather roundabout method is needed because commercially available colour film is not designed for long exposures to low-light levels (it usually turns a murky green!).

These wide-band filters are useful for studying stars, which do not produce radiation of a single wavelength but emit light over a wide range of wavelengths – the whole rainbow spectrum of colours. A gas produces a continuous spectrum of this kind when it is dense, so that the atoms are comparatively close together. When atoms are relatively isolated from one another, however, as they are in a nebula, they emit light of just a few, very specific wavelengths. Each element produces its own particular wavelengths, which act as its spectral 'fingerprint'. Hydrogen emits intense red light at a wavelength of precisely 656.28 nanometres and also fainter blue-green and blue

light (at 486 and 434 nanometres). Oxygen produces intense green light at 501 nanometres wavelength.

An astronomer can take advantage of these specific wavelengths. If investigating faint nebulosities, a 'narrow band' filter can be put in front of the telescope to pass only light of a particular wavelength. A common ploy is to study the bright red hydrogen light with a filter that passes only light whose wavelength is within five nanometres of the theoretical wavelength of the hydrogen light. Such filters have revealed huge faint nebulosities filling the constellation of Orion, and hundreds of dim loops of gas in a nearby galaxy, M33 (Fig. 10.20).

The techniques described so far all produce a two-dimensional *image* of some object in the sky. But this is only one way to investigate the Universe – and an image is of little use in the study of a star so distant that it appears as no more than a point of light in even the largest telescope. The second major spearhead is *spectroscopy*. Here the telescope is used to gather light from just one object – a star perhaps, or a small region of a nebula or galaxy. This light is passed into a spectrograph, which spreads it out as a spectrum of wavelengths – a band running from the blue-violet wavelengths at one end to red at the other, like a precise section through a rainbow. Within this band, the emissions from particular elements fall at their characteristic wavelengths, as lines cutting across the spectral colours (Fig. 3.1). The lines from the glowing gas of a nebula are always bright *emission lines*; but the lines in a star's spectrum are generally dark *absorption lines* – silhouettes against the star's bright continuous spectrum of colours.

Spectroscopy thus can identify the elements that are present in a star. A detailed analysis of the intensities of lines reveals the relative abundances of the different elements in the star, the star's temperature, and whether it has a strong magnetic field.

By studying the precise wavelengths of the lines in a spectrum, it is possible to determine the speed at which a star is travelling towards or away from us. If a star or galaxy is moving away from us, its radiation is slightly stretched out and all the wavelengths are very slightly longer than they should be; if it is coming towards us, the wavelengths are bunched together and shortened (Fig. 3.13). This distortion of wavelengths is known as the *Doppler effect*; it is exactly the same effect which raises the pitch of an ambulance siren as it comes towards us, and abruptly drops the pitch as the ambulance passes us and

moves away at speed. Most stars in our Galaxy travel at speeds of around 20 kilometres per second (70 000 kph) – very fast by everyday standards, but only enough to shift their spectral lines by less than a tenth of a nanometre. One very odd 'star' called SS 433 (Figs. 6.34–6.38), however, is ejecting streams of gas at 81 000 kilometres per second, so rapidly that the spectral lines appear in totally the wrong place in the spectrum – and they move up and down the spectrum as the gas streams change direction.

Beyond our Galaxy, the other galaxies show Doppler shifts in their spectra. All the distant galaxies are moving away from us, carried by the general expansion of the Universe so that the more distant ones are moving faster. Their motion away from us stretches their radiation to longer wavelengths, with the result that any particular spectral line is moved towards the red end of the spectrum, and generally the Doppler shift of a galaxy is called its *redshift*. Knowing how rapidly the Universe is expanding, astronomers can convert the measured redshift of a galaxy into its actual distance from us. For the most distant detectable objects of all, the exploding disrupted centres of

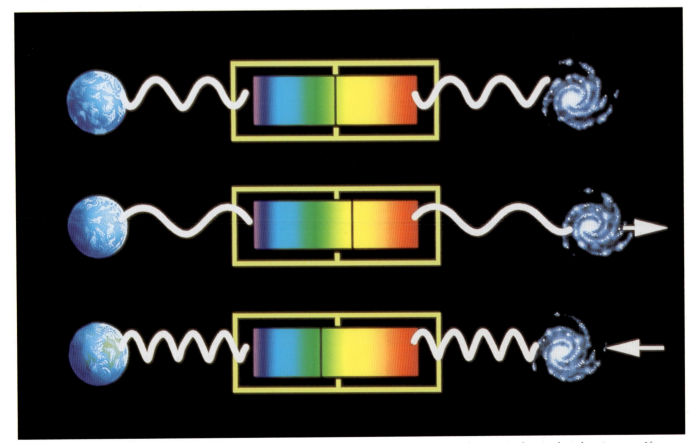

Fig. 3.13 *Lines in the spectra of stars and galaxies reveal how fast they are moving towards or away from us. At the top, the galaxy is at rest. If a galaxy is receding (middle), its wavelengths are 'stretched' so that its spectral lines are shifted towards the red end. The approaching galaxy's wavelengths (bottom) are compressed, shifting its lines towards the blue.*

Fig. 3.14 *The Hubble Space Telescope is deployed from the space shuttle Discovery on 25 April 1990. Two unfurled solar panels (gold) provide electrical power. The shutter over the near end of the tube prevents sunlight from inadvertently shining into the telescope: it opens to give Hubble unprecedently sharp views of the Cosmos.*

far-off galaxies which are called quasars, the redshift is the only guide we have to their distances.

With the quasars and the most distant galaxies, we reach – quite literally – the frontiers of astronomical discovery. And astronomers need new tools to probe further. Optical astronomers are currently advancing on several fronts.

The first is the quest to see more detail, to beat down the blurring effects of the Earth's turbulent atmosphere. In theory, the larger the telescope mirror, the more detail it can resolve. But the atmosphere spreads out the image of a single star to around one arcsecond in size. In the case of an extended object, like a planet, nebula or galaxy, details smaller than one arcsecond are smeared out to the point of invisibility. As a result, even the largest telescopes on Earth can see details no finer than an average amateur astronomer's backyard telescope!

If enough money is available, there is one easy solution: put your telescope above the atmosphere altogether. This is the philosophy behind the Hubble Space Telescope (Fig. 3.14). Named after the American astronomer Edwin Hubble, who

pioneered research into galaxies in the 1920s, this $1.4 billion telescope was launched into a 600-kilometre high orbit by the space shuttle in 1990.

The Hubble Space Telescope has a main mirror 2.4 metres in diameter, which means it can resolve details as small as 0.1 arcseconds. Astronomers have a choice of two CCD cameras on board. The Wide Field/Planetary Camera is the workhorse instrument, and is able to view a region up to 150 arcseconds across. The Faint Object Camera sees a smaller region, but in finer detail. Hubble's instruments at launch also included two spectrographs and a photometer for measuring rapid changes in the brightness of a star or quasar. This first set of instruments covers a wide range of wavelength, from the red end of the visible spectrum right down to 115 nanometres in the ultraviolet.

Soon after launch, the Hubble astronomers discovered that the main mirror had been polished to slightly the wrong shape. The images suffered from spherical aberration, giving each star a halo of light. During a space shuttle mission in December 1993, however, astronauts

Fig. 3.15 *Hubble's spectacular views of nebulae and galaxies have astounded astronomers. The Cartwheel galaxy (left) was previously known only as a bright ring surrounding a fuzzy hub: Hubble has revealed intricate detail. The straight spokes and the young stars studded round the rim show that the Cartwheel acquired its unique shape only recently, probably when one of the small companions ran right through a normal spiral galaxy.*

inserted a small set of mirrors that would intercept the light from the main mirror and correct it before it strikes the instruments. This 'fix' worked perfectly, and Hubble began to return images fully as sharp as the original specification.

Meanwhile, other astronomers are attempting the much cheaper option of sharpening up the view as seen from the ground – aided by 'Star Wars' technology designed for the opposite task of keeping a laser beam narrow as it heads up through the atmosphere towards an incoming missile.

If we could see the waves of light from a distant star or galaxy as they cross space, they would look like the waves in the middle of an ocean – straight parallel lines progressing across the emptiness. When these straight wavefronts enter the Earth's atmosphere, irregularities in the air bend different parts of the wave in different directions. When these corrugated wavefronts are reflected by a telescope mirror, they do not reach precisely the same focus and so produce a blurred image.

In the 1980s, astronomers realised that they could straighten out a distorted wavefront with a distorting

mirror. In the method of *adaptive optics*, the light from the telescope's main mirror falls onto a small flexible mirror. Dozens of small actuators behind this mirror distort its surface to exactly the shape of the incoming light-waves, so straightening out the wavefronts.

Because the atmosphere is constantly changing, the shape of the flexible mirror must be continually altered. And that means knowing – at every instant – the distortion that the atmosphere is introducing. The simplest way is to observe a bright star near to the faint galaxy or quasar you are really interested in. By intercepting some of the abundant starlight and knowing that the star should appear as a single point of light, a computer can quickly measure the distortion and so feed the appropriate corrections to the flexible mirror. Since the galaxy's light travels through the same region of atmosphere, the technique sharpens it as well.

But nature is not kind enough to provide a bright star near every interesting astronomical object. One alternative is to create an 'artificial star' by shining a powerful laser up into the atmosphere. The laser light stimulates a small

Fig. 3.16 *Laser beams shoot through the New Mexico sky, in an experiment that may let Earth-bound telescope compete in clarity with the Hubble Space Telescope. The light from the green laser is absorbed in a small patch high in the atmosphere above the Starfire Optical Range. From the big dome at the base of the beam, it looks like a tiny artificial star which can be positioned near to any real star, nebula or galaxy. A computer attached to the telescope works out how the light from this artificial star is distorted as it comes back down through the atmosphere, and shapes a flexible mirror to focus the light back into a point. As a result, the flexible mirror will also compensate for the distortions in light from anything else in the same direction, so sharpening up the view of distant astronomical objects.*

patch of sodium atoms, which occur naturally in a layer 100 kilometres above sea level. It looks like a faint star (Fig. 3.16). The light coming back from this high-altitude 'artificial star' has to traverse the lower atmosphere, just like light from an astronomical object, so astronomers can use it as the bright star needed for adaptive optics. And the great advantage is that they can position the artificial star wherever in the sky they like.

A few astronomers are trying to go for the ultimate in resolution, with arrays of optical telescopes linked together to simulate the effect of a single telescope hundreds of metres across. This *optical interferometry* is similar to the technique of aperture synthesis widely used in radio astronomy (Chapter 7). Prototype optical interferometers in California, Australia and Britain have shown the technique can work, but it will be well into the next century before we are regularly seeing images with details of a thousandth of an arcsecond.

The major drive in optical astronomy now is once again for larger telescopes. The highly-efficient CCD has meant that existing telescopes have become much more powerful. Fitted with a modern CCD, the Hale Telescope is 200 times more sensitive than when it used 1940s photographic plates – equivalent to increasing its diameter 14 times.

With CCDs now picking up virtually all the available light, the only way to increased sensitivity is with larger mirrors. But there is a fundamental problem: a large thick

mirror is impossibly heavy, while a large thin mirror will bend under its own weight as the telescope is tipped up.

Astronomers have devised three possible solutions. Roger Angel, at the University of Arizona, uses honeybee technology to make thick mirrors that are three-quarters air. Each mirror has a curved front face and flat back plate, held together by a honeycomb lattice of glass (Fig. 3.17). Angel has cast a 6.5-metre mirror, and is gearing up to produce mirrors 8 metres across.

Astronomers at the European Southern Observatory have pioneered the opposite approach. Their New Technology Telescope has a very thin mirror, held up by adjustable supports. A computer works out how the mirror deforms under its own weight, and commands the supports to push it back to the correct shape. This technique, *active optics*, is now being applied to a large Japanese telescope, with an 8.3-metre mirror, and to the European Southern Observatory's Very Large Telescope. The latter will consist of four linked telescopes, each with a mirror 8.2 metres in diameter, on the peak of Cerro Paranal in Chile. Around the year 2000, it will become the most powerful telescope in the world.

In the meantime, that accolade belongs to a telescope with an even more audacious design. The Keck Telescope (Fig. 3.18), on the summit of Mauna Kea, has a mirror 10 metres across – too big to be made in one piece. So designer Jerry Nelson has built the mirror from

36 hexagonal segments, each 1.8 metres across. The segments are small enough not to bend, and the onus is on a hundred computer-controlled supports which constantly push and pull the segments so that they form a continuous surface of precisely the correct shape. This unorthodox design has proved so successful that a twin telescope, Keck II, has been built at the other end of the control building for the first Keck Telescope.

These giant new instruments will increase the world's total area of telescope mirror by a factor of four between 1990 and 2000. The Very Large Telescope on its own will collect more light than all the world's telescopes together in the 1980s.

Optical astronomy may be the oldest branch of astronomical research – with its roots stretching back thousands of years into prehistory – but it is also at the forefront of science today. With the precision of the Keck Telescope, the complexity of the Hubble Space Telescope and the computer wizardry involved in analysing CCD images, we are at the frontiers of technology too.

Many youngsters are spurred to a career in astronomy by the thrill of seeing a dark star-studded sky. And although, as professional astronomers, the naked eye or small telescope may be superseded by huge steel and glass reflectors with a silicon chip 'retina', there is still as much fascination – and endless information on the Universe – to be gained from the optical radiation which reaches us from space.

4 starbirth

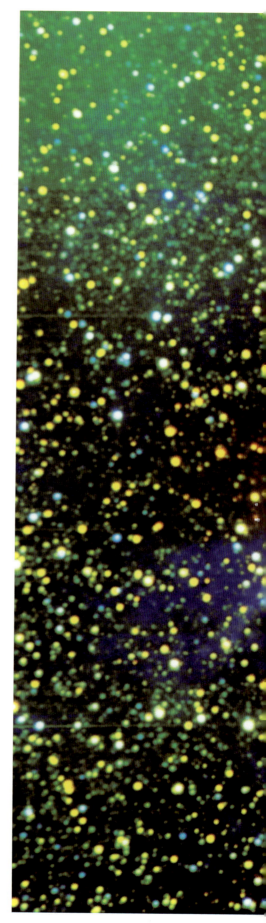

OUR EYES CAN detect about 6 000 stars while large optical telescopes show us thousands of millions of fainter and more distant stars. These stars differ from one another in mass, size, brightness and temperature. But almost all have one thing in common: they are in the prime of life.

There must, however, be some stars forming and others dying (Chapter 6). At the present time this happens at a rate of about ten stars born and ten dying in our Galaxy each year. New stars are born when the very tenuous gas in between the existing stars is compressed, either by one of the Galaxy's spiral arms or by the shock wave from the explosion of a supernova. But it is very difficult to see starbirth in progress. The interstellar gas contains small grains of dust, like cosmic soot, and where the gas is compressed the dust builds up until it almost totally blocks the light coming from the 'star nursery' within. The dust absorbs and scatters light very effectively because the grains are about the same size as the wavelength of visible light. The longer wavelengths of infrared and radio, however, can penetrate the dust quite easily, and reveal star formation within the clouds.

Radio waves show up various stages of the birth process, in 'lines' of particular wavelengths from molecules. These groups of atoms can only exist in dense, dusty clouds where they are protected from disruptive ultraviolet radiation, so surveys of the sky for radio emission from molecules reveal the dense clouds – *molecular clouds* – where stars are about to form. Detailed surveys have revealed dense clumps of gas within the molecular clouds. The gas forming each of these 'cores' has enough gravitational pull to draw itself inwards and create a new star.

We must turn to infrared telescopes to catch stars as they are born. Astronomers cannot hope to see a single star 'switch on'. Instead, they search the individual cores in molecular clouds to find clumps of gas at different stages on the route from dense core to genuine star.

In the molecular cloud, the temperature is only 10 K, and the dust emits infrared of about 300 micrometres wavelength. As a core shrinks in size, it becomes hotter and brighter. When a core with the Sun's mass collapses, the squeeze of its own gravity makes it shine much more brilliantly than the Sun itself.

At this stage, it is still a *protostar:* with no nuclear reactions, its energy comes from gravitational collapse. Radio observations often reveal a surrounding disc of gas and dust – which may condense into planets – spewing out jets of hot gas.

As the protostar continues to shrink, the central temperature reaches 10 000 000 K. Hydrogen nuclei start to fuse together to form helium. This nuclear reactor will supply a steady stream of energy for millions, or even billions, of years. The protostar has become a star. But the moment of starbirth is an understated event. At first, the nuclear reactions are overshadowed by the energy from the star's continuing contraction, so starbirth is signalled by only a small increase in brightness.

The radiation from a hot young star can heat the surrounding gas, causing it to shine brightly at radio wavelengths. Its energy soon strips away the surrounding dark natal cloud, so the hot gases become visible to optical telescopes, as a nebula.

With the techniques of the new astronomy, astronomers have found that starbirth can be a self-perpetuating process. An infrared view of the nebula M17 (**Fig. 4.1**), lying 5 000 light years from the Sun, provides a perfect example. It was discovered by a French astronomer, Philippe de Cheseaux, in 1746. He saw only a small elongated nebula – the bluish region in the centre here – surrounding a cluster of young stars.

But dark dust clouds are hiding the main scene of action from optical telescopes. Fig. 4.1 shows a brilliant bar of hot gas, newly-born stars and protostars to the lower right of the optical nebula. They make M17 one of the brightest objects in the whole sky when viewed at infrared wavelengths. To the right of the bar, the molecular cloud is so thick that it hides the background stars even at penetrating infrared wavelengths.

This is a scene of continuing starbirth. The young stars in the optical nebula are pumping out energy which compresses the edge of the molecular cloud, creating the young stars and protostars that dominate the infrared view. In turn, they will squeeze neighbouring regions of the molecular cloud, inducing the gas and dust further in to collapse into stars. Eventually, all the gas and dust in the giant dark cloud will condense. A million new stars will be born.

4.1 M17. *Infrared, 1.2 μm (blue), 1.65 μm (green), 2.2 μm (red), 3.8 m Mayall Telescope*

Orion region

THE STARS OF Orion (the hunter) make up the most brilliant of all the constellations, containing one-tenth of the seventy brightest stars in the sky. Bright red Betelgeuse forms one of the hunter's shoulders; below, a line of three stars makes up his Belt, with a nebulous Sword hanging from it; and two lower stars make the bottom of his tunic – the one on the right being Rigel, the seventh brightest star in the sky. The figure of Orion shows clearly on the blue-filtered photograph (**Fig. 4.2**), although the filter dims the red star Betelgeuse (top left).

In the direction of Orion, we are looking at the major region of starbirth in our region of the Galaxy, the local spiral arm. Spiral arms are the major sites for star formation, so throughout Orion we see young stars, including the hot, luminous stars which have only a short lifetime. These O and B type stars have temperatures of 20 000 K and above, and they shine bluish white. (Betelgeuse has already almost completed its short span, and is now a red giant star nearing the end of its life.)

Brilliant Rigel, 800 light years away, is the nearest of Orion's B stars. Twice as far away, we come to a group which is extremely young. These stars of the Orion OB Association include the stars of the Sword, the Belt and the fainter stars surrounding them and extending to the upper right of the Belt. The stars have condensed from a single large gas cloud. Judging by the stars' ages, star formation started at the top right of the cloud some 12 million years ago, and has gradually spread along it. The stars of the Belt region formed 8 million years ago, and the youngest in the Sword are only 2 million years old.

The three stars of Orion's Belt are more massive and produce more energy than Rigel. But they are so hot that most of their energy comes out at ultraviolet wavelengths, and our eyes do not see them in their full glory.

The Belt stars are 30 times heavier than the Sun, and outshine our local star some 100 000 times. With this immense output of energy, they are surrounded by huge hot atmospheres, giant versions of the corona that surrounds our Sun (Fig. 2.14). This hot gas is a powerful source of X-rays.

As a result, the young massive stars of the Belt show up clearly in an X-ray view of the Orion region (**Fig. 4.3**) – upper left – along with the nebula (bottom left). The scene is peppered with other superhot stars that hardly appear on the optical photograph.

4.2 *Optical, blue light, 8 cm camera lens, Harvard College Observatory at Agassiz*

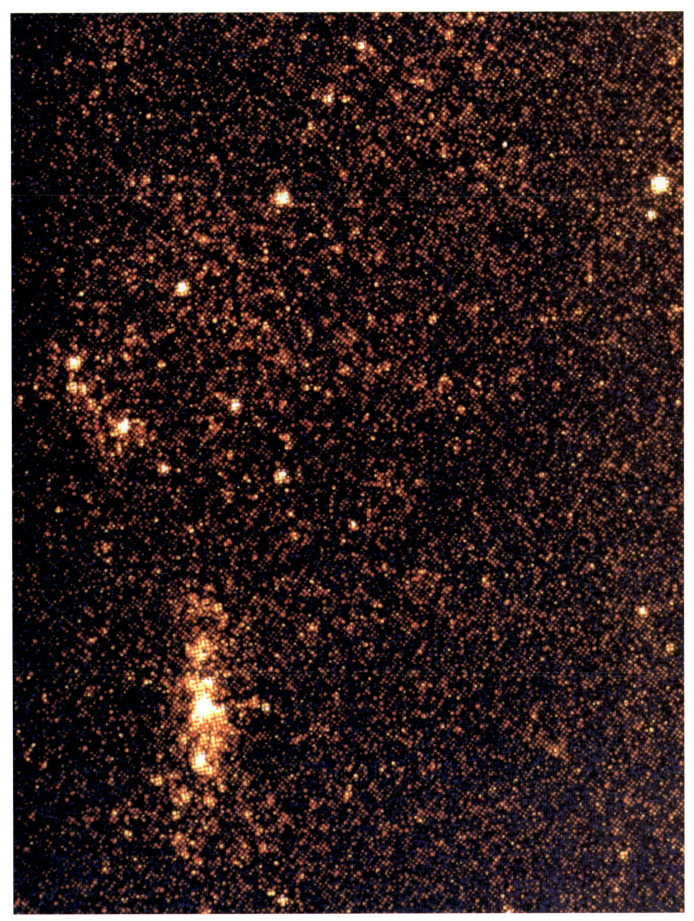

4.3 *X-ray, 0.6–10 nm, Position Sensitive Proportional Counter, Rosat*

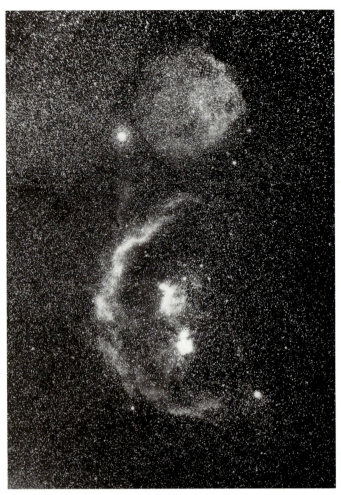

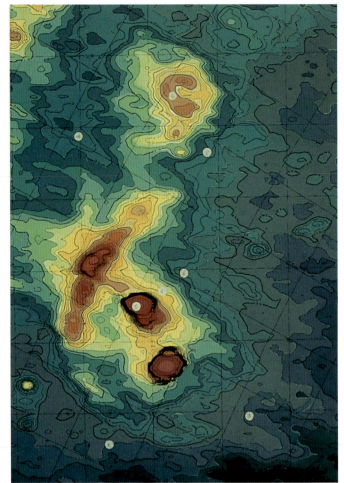

4.4 *Optical, red light, photographically amplified, camera mounted piggyback on 0.4 m telescope, Siding Spring*

4.5 *Radio, 21 cm continuum, 25 m Stockert Telescope, Bonn University*

To our eyes, Orion appears as a sprinkling of crisp stars on a black velvet background. But this region of the Galaxy is filled with astonishing loops and whorls of gas and dust.

In 1894, the American astronomer E. E. Barnard photographed Orion and discovered a huge faint loop of glowing gas. **Fig. 4.4** is a recent photograph of Barnard's Loop. Approximately a semicircle, it curves from above the Belt southwards and around towards Rigel. The Loop is made of hot gas, and shines brightly in the red light from hydrogen. To bring out details, Fig. 4.4 was photographed through a red filter. This filter also accentuates the red star Betelgeuse (top left) while dimming the blue-white stars of the Belt.

Barnard's Loop is 300 light years across, and has been expanding for two million years. It is a bubble of gas blown by the young and brilliant stars in Orion's Sword.

A circle of glowing gas appears in Fig. 4.4 around the 'head' of Orion, while two bright nebulae are visible in the middle of Barnard's Loop. The lower is the famous Orion Nebula, while the upper is the hot gas surrounding the Horsehead Nebula.

A radio map (**Fig. 4.5**) shows longer-wavelength emission from the same hot gas. The background sky is coded blue, and successively brighter radio emission is shown in yellow, orange and red. Stars are not detected, so the main stars of Orion have been superimposed for reference.

The two brilliant nebulae in Orion again show clearly, along with Barnard's Loop and the Orion's Head Nebula. But the radio view has two advantages. First, it does not suffer from the confusing number of stars that appear on photographs. And the radio view is not obscured by dark clouds. Compare, for example, the region between the left-hand star of the Belt and Barnard's Loop. The radio view (Fig. 4.5) shows a band of hot gas, which is hidden in the optical photograph (Fig. 4.4) by a dark cloud of dust.

The Infrared Astronomical Satellite (IRAS) has literally brought to light whole festoons of dusty clouds in Orion. At a temperature of a few hundred K or less, they emit strongly at infrared wavelengths.

Fig. 4.6 is IRAS's view of Orion, with colour coding showing temperature. Blue corresponds to several thousand K; yellow to several hundred degrees; and red to a few tens of degrees above absolute zero. The stars visible here are all red giants:

because of the coding, they appear blue in this false colour view. Betelgeuse is the brightest star, towards the top left of the image.

The two brilliant white nebulae are again the Orion Nebula and the region of the Horsehead. In these regions, warm clouds of dust are mixed in with the hot hydrogen gas seen in Figs. 4.4 and 4.5.

In the IRAS view, notice the ring of dust at the top. It surrounds the head of Orion, and looks at first glance like the circle of glowing hydrogen visible in the two previous images. In fact, the ring of dust is larger, and encloses the hot gas. Radiation from the central star is heating the gas in its vicinity to incandescence, but further out the radiation weakens to the point where it can only warm the dust.

Otherwise, there's little correspondence between the radio and infrared views. The hot gas in Barnard's Loop, for example, is too tenuous to show in Fig. 4.6, while warm dust clouds do not emit radio waves. But each has an equally important story to tell. The radio view reveals where hot bright stars have formed recently, while the IRAS image shows where stars are being born right now.

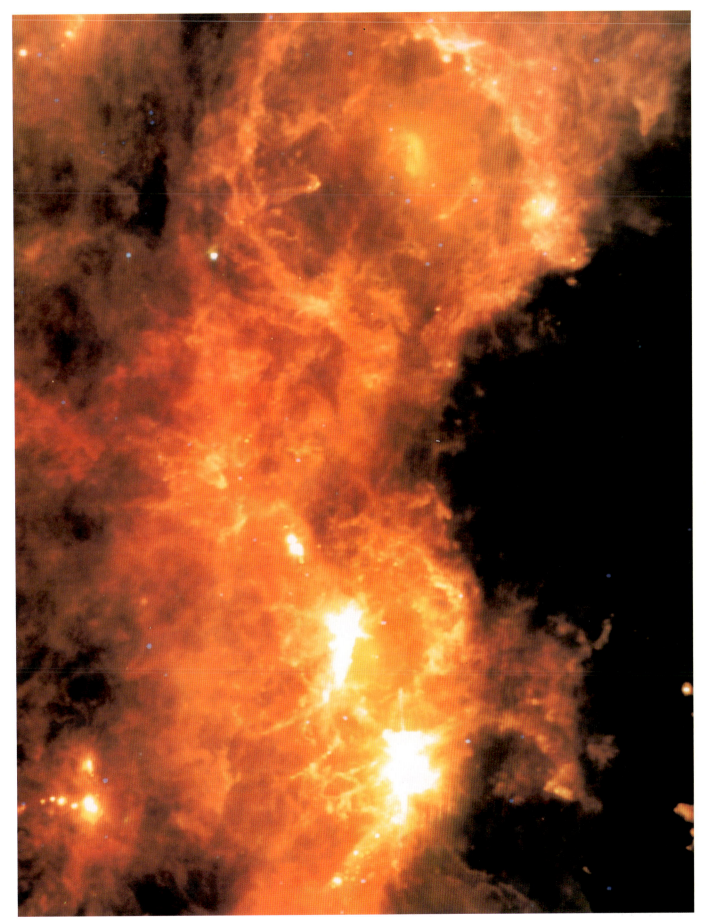

4.6 *Infrared, 12 μm (blue), 60 μm (green), 100 μm (red), IRAS*

4.7 *Optical, true colour, 1.2 m UK Schmidt Telescope*

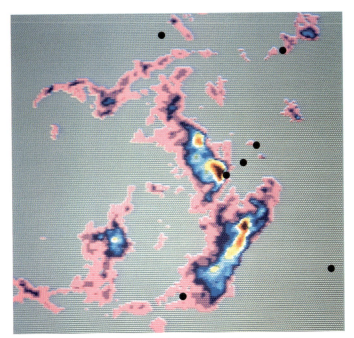

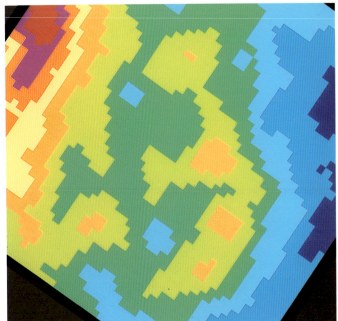

4.8 *Radio, 2.6 mm carbon monoxide line, 1.2 m Sky Survey Telescope, Columbia University, New York*

4.9 *Gamma ray, 0.000 002–0.000 01 nm, EGRET, Compton Gamma Ray Observatory*

Orion still contains vast reservoirs of gas and dust, the raw materials of stars still to be born. In the blackness of space, these dense 'molecular clouds' generally appear dark and invisible to optical telescopes. But sometimes we are lucky enough to see one silhouetted against a bright background – and a beautiful sight it can make. In **Fig. 4.7**, a dark cloud is filling the left-hand side, obscuring the stars beyond. Along its boundary runs a rim of gaudily glowing gases which provides a backdrop for one of the best loved of astronomical objects, the Horsehead Nebula. This view is printed with North upwards, so the nebula appears to be lying nose-up: turn the page through 90 degrees for a more conventional view of the cosmic dark horse.

The star at the top is the left-hand star of Orion's belt, Alnitak. But it merely lies in the foreground, and has no role to play in lighting up the scene. The illumination comes from sigma Orionis, in the lower right of Fig. 4.7. It is a blue-white main sequence star, at a temperature of 30 000 K. Its ultraviolet radiation is heating up the edge of the dark cloud, and 'boiling away' its gases. They glow with the red light of hot hydrogen, to form a nebula catalogued as IC 434. It is 30 light years in total size, and contains as much matter as 250 Suns.

The famous Horsehead is a dark cloud silhouetted against IC 434. Although it looks small in Fig. 4.7, in reality it measures 3 light years from nose to mane – almost the distance from the Sun to the nearest star. The 'head' of the horse was originally a particularly dense clump of gas and dust within the dark cloud. It has proved more resistant to the erosion by sigma Orionis, so it now stands proud of the edge of the cloud.

The bright blue patches are regions where starlight is scattered by the tiny dust particles in the cloud, while the nebula at the top, NGC 2024, is a partially hidden region of starbirth that generates most of the radio emission from this region.

With an optical telescope it's not so easy to make out the other boundaries of the huge dark cloud that the cosmic horse is poking his head out of. But the cloud is thick with molecules, each broadcasting its own characteristic radio wavelengths. When viewed at these wavelengths, the whole dark cloud is ablaze with radiation.

Fig. 4.8 shows the whole constellation of Orion as seen at the wavelength emitted by carbon monoxide molecules – not the commonest molecule in the clouds, but one that broadcasts very powerfully. Here the background sky is coded turquoise; feebly-emitting regions are pink and successively more intense regions blue, yellow and brown. The superimposed stars reveal the region of sky mapped here. Orion itself contains two elongated molecular clouds, while a smaller cloud lies to the left in the constellation Monoceros.

The upper cloud in Orion is the parent of the Horsehead Nebula. The sharp edge to the cloud, where it is being whittled away by sigma Orionis, shows up as the abrupt change in colour at the right-hand side, from brown to the background pink. The Horsehead Nebula itself is too small to show up on this scale: it would be less than a millimetre across.

This molecular cloud is 150 light years long, and contains enough matter to form over 100 000 new stars. Its twin, in the southern part of Orion, is slightly more massive. The brightest point in this lower cloud lies behind the famous Orion

Nebula, and is shown in more detail in Fig. 4.18. Originally, both clouds extended towards the right of Fig. 4.8, and may have joined here. But the gas in this region has already condensed into the stars of the Orion OB Association. Star formation is now proceeding down and to the left, eating into the two surviving clouds.

The extent of these clouds was unknown before the carbon monoxide observations, as were the many fainter streamers of gas. Some are extremely elongated: up to 500 light years long but only 20 light years thick.

The clouds also show up when we make observations at the other end of the spectrum of electromagnetic radiation. **Fig. 4.9** shows the same region as Fig. 4.8, but viewed at gamma-ray wavelengths. Colour coding shows brightness, from purple for the most intense, down through shades of brown, orange and green to blue for the faintest regions. The Milky Way runs across the top left, while Orion's two molecular clouds stand out clearly, though not in the same detail.

These gamma rays are emitted when high-speed particles – cosmic rays – strike the atoms and molecules in space. So the intensity of gamma-ray emission also indicates the density of the invisible clouds in Orion. Despite the different emission process, and a wavelength a million million times shorter, the gamma-ray image reveals a distribution of gas clouds very similar to that seen in the carbon monoxide image.

The images can only be identical, however, if the flux of cosmic rays is the same all over this large region of space. Astronomers are now comparing carbon monoxide and gamma ray images of Orion in detail, to map out the distribution of cosmic rays in a region over a thousand light years away from home.

4.10 *Optical, true colour, Wide Field/Planetary Camera 2, Hubble Space Telescope*

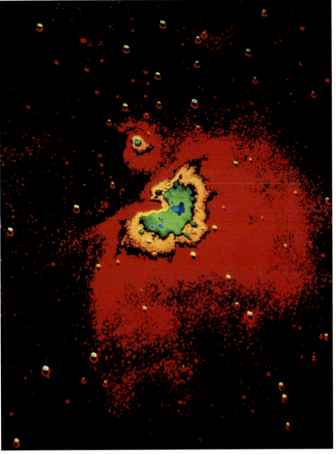

4.11 *Optical, intensity coded, 0.2 m reflector*

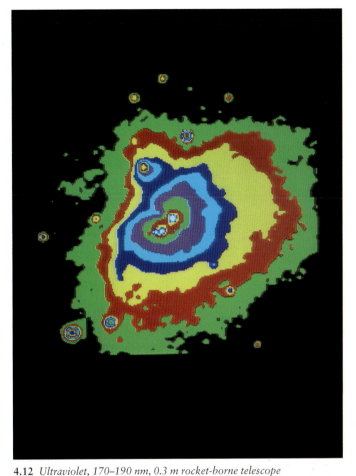

4.12 *Ultraviolet, 170–190 nm, 0.3 m rocket-borne telescope*

Orion Nebula

THE FAN-SHAPED TRACERIES of the Orion Nebula consist of hot gases spread out over 15 light years of space. Although the Nebula lies 1600 light years away, it is large enough to appear the size of the full Moon (30 arcminutes across) in our skies.

In close-up, the Orion Nebula is stunning (**Fig. 4.10**). This true-colour image from the Hubble Space Telescope is a thousand times crisper than our naked eyes can see. It shows how the nebula's heart would appear from a planet orbiting a star in the outskirts of the Orion Nebula itself. Yet even the closest optical view cannot reveal all the nebula's hidden secrets.

The Orion Nebula is also called M42, as the forty-second entry in Charles Messier's 1784 catalogue of nebulous objects. A small, round nebula just above is known as M43. Both nebulae are lit up by ultraviolet radiation from hot, newly-born stars at their centres. As a result, each nebula is very much brighter at the centre than at the edge, making it difficult to photograph the whole nebula in a single exposure. American astronomer Fred Espenak has taken an ordinary long exposure photograph, and coded the different levels of brightness in different colours to create an isophote picture (**Fig. 4.11**). The most brilliant regions, around the central stars,

are shown blue, while the surrounding slightly fainter gas is coded green. This is roughly the extent of the nebula as seen through a small telescope. Fainter levels are shown yellow and red.

David Malin, of the Anglo-Australian Observatory, has developed a very different technique which suppresses this range in brightness, so that fine details show up in ordinary photographic form from centre to edge. The unsharp-masked photograph (**Fig. 4.13**, overleaf) has been made by printing the negative through an unfocused positive copy of itself to reduce the brightness range without losing detail. Three separate exposures with different coloured filters have been unsharp-masked to show the nebula's true colours.

The weaker radiation reaching the outer regions of the nebula causes hydrogen to glow in its characteristic red light, at a wavelength of 656 nanometres. Towards the hot stars at the centre, more energetic short wavelength ultraviolet lights up many other atoms. The strongest emission comes from oxygen atoms which have lost two electrons, and shine at green wavelengths.

Malin's photograph shows up the stars responsible for lighting up the nebulae. In M43, it is the single star in the centre, NU Orionis, a star at a temperature of about 25 000 K and shining as brightly as several thousand Suns. The Orion Nebula itself,

M42, is centred on a small group of hot stars called the Trapezium.

Photographs of the nebula at visible wavelengths show mainly the distribution of hot gas. But M42 also contains much of its original dust. This reflects shorter wavelengths better than longer, and Malin's photograph shows vividly the bluish-grey dust lanes sweeping forward around the top left edge of the blister burnt by the Trapezium in the dark dense cloud behind.

The ultraviolet photograph (**Fig. 4.12**) is dominated by radiation from dust. It was taken by a 31-centimetre telescope with a microchannel plate detector which was lofted above the Earth's ozone layer by an Astrobee rocket. The image (originally recorded on photographic film) has been converted to an isophotal map, with the brightness levels running down from pale blue for the bright centre, through dark blue, yellow, red and green to purple; and then through a second cycle of these colours.

In the ultraviolet, radiation reflected from dust reduces the contrasts in the nebulae: the optically dark dust lane between M42 and M43 actually glows (at the 'dark blue' level), while the 'bite' out of the left of M42 is hardly noticeable. Because dust is spread all around M42 and M43, the nebulae together appear more circular in the ultraviolet picture.

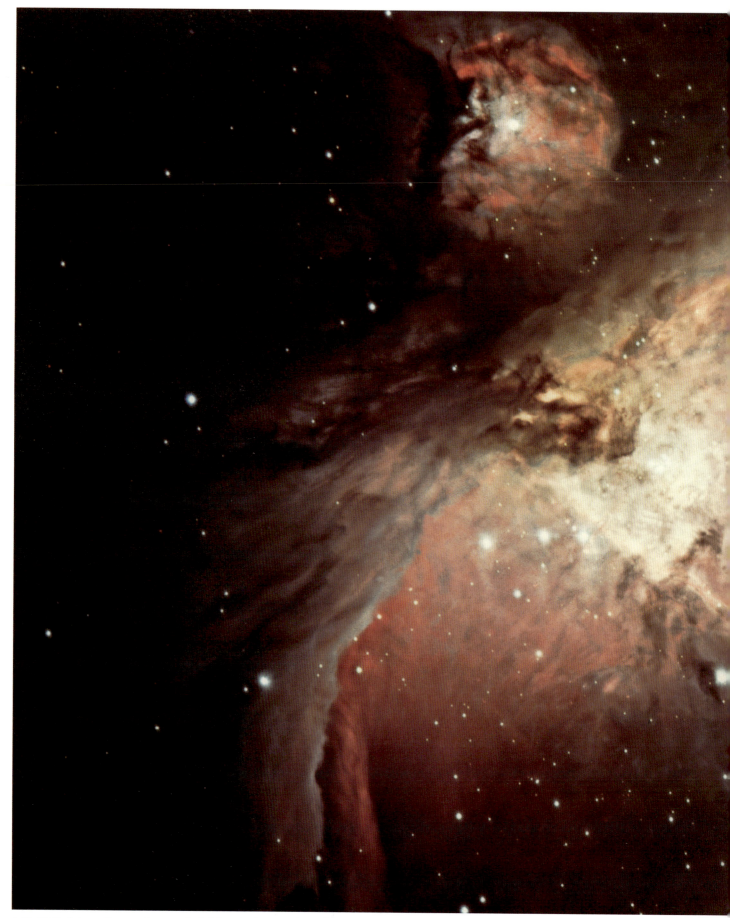

4.13 *Optical, true colour, unsharp masked, 3.9 m Anglo-Australian Telescope*

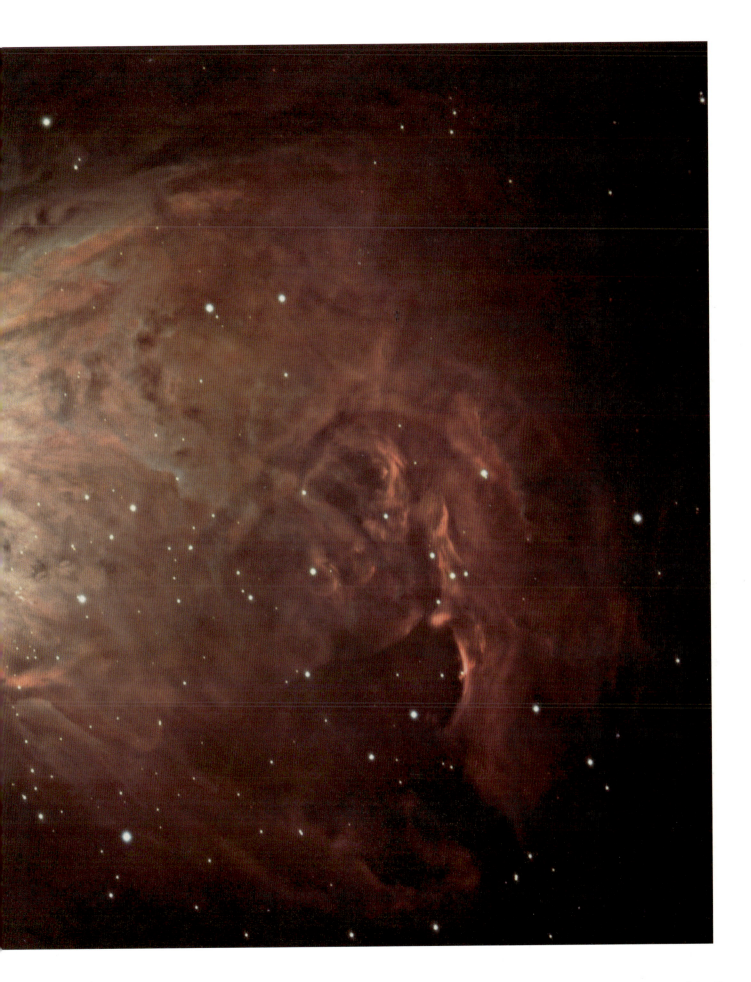

The newly-born stars in the Orion Nebula show up prominently at X-ray wavelengths: the Einstein X-ray Observatory found over a hundred. Its view of the central six arcminutes is here (**Fig. 4.14**) superimposed as a contour map on an optical photograph from the Lick Observatory. Each 'peak' or closed contour line is a source detected by Einstein's High Resolution Imager. Almost all have been identified with stars, labelled here either by their conventional letter designation or by their 'π' number from P. P. Parenago's catalogue of stars in the nebula .

The lower picture (**Fig. 4.15**) is the central region of this Einstein view, to twice the scale. Here the X-ray intensity has been converted to brightness, and only the most intense sources show. The brightest X-ray star, theta-1C, is the southernmost (lowest) of the four stars in the Trapezium cluster; theta-2A is the right-hand of the three stars visible on the optical photograph forming a line at the lower left of the Lick photograph. 'MT' and 'LQ' are relatively faint at optical wavelengths; they are two of the many young variable stars, whose brightness changes erratically as they stabilise into 'main sequence' stars.

The central star theta-1C is far more powerful than the other three Trapezium stars, which do not show up at all. The X-ray view underlines the fact that although the four Trapezium stars look rather similar at visible wavelengths, theta-1C is by far the most impressive – and important. With a mass of about 20 Suns, it is approximately twice as heavy as any of the others; it has a surface temperature some 10 000 degrees hotter, at 40 000 K; and it emits almost ten times as much radiation at visible and ultraviolet wavelengths. Its searing surface is surrounded by an even hotter corona (outer atmosphere). At a temperature of 20 million K, the corona shines a million times more brightly than does the Sun's corona in X-rays.

The ultraviolet radiation from theta-1C is almost solely responsible for heating and lighting up the whole of the Orion Nebula. Without this one star, the nebula would be invisible to the naked eye.

The hot gas in the nebula also emits radio waves copiously. **Fig. 4.16** is a radio view, to roughly the same scale as Fig. 4.14. It is colour coded so that the fainter regions appear blue, with successively brighter parts yellow and red. According to the radio observations, this central region of the nebula contains ten Sun-masses of gas spread over two light years; the extended outskirts contain perhaps ten times as much again.

Fig. 4.16 shows that the gas forms a network of tangled filaments. The most elongated filament, to the upper left, is one and a half light years long. The radio emission in this region corresponds to the

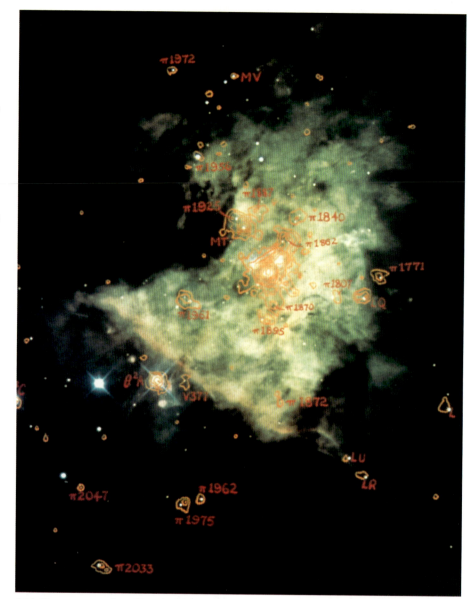

4.14 *X-ray contours from Einstein Observatory, overlaid on true-colour optical photograph, 3 m reflector, Lick Observatory*

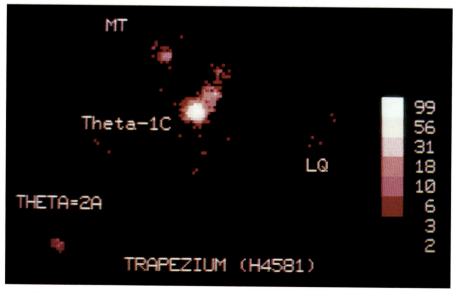

4.15 *X-ray, 0.4–8 nm, High Resolution Imager, Einstein Observatory*

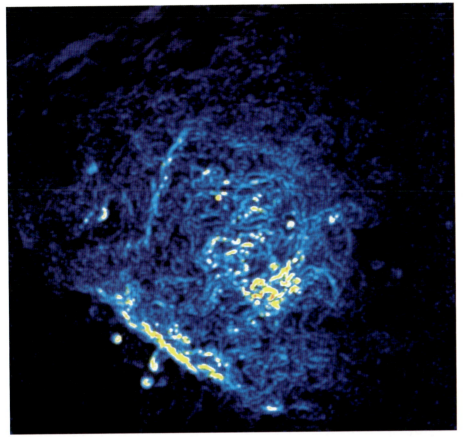

4.16 *Radio, 21 cm continuum, Very Large Array*

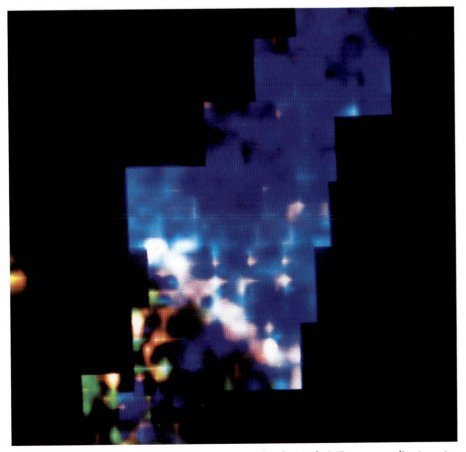

4.17 *Ultraviolet, 191 nm carbon line (blue), 233 nm carbon line (red), 247 nm oxygen line (green), 0.45 m reflector, International Ultraviolet Explorer*

dark 'bite' out of the nebula seen in optical pictures. The bite, then, is not caused by a lack of hot gas, but is due to a flap of the surrounding dark cloud that blocks visible light.

The tangled filaments probably mark the edges of dozens of separate bubbles of hot gas, blown by the O and B stars that throng the centre of the nebula. The brightest region of filaments lies just to the lower right of the Trapezium, only 0.1 light years from theta-1C itself.

The diagonal line at the lower left of Fig. 4.16 is the 'bar' that is also prominent in optical photographs. Despite the name, it is not actually spindle-shaped, and appears bright merely by an effect of perspective. The whole Orion Nebula is a thin saucer-shaped layer of glowing gas, where the surface of a dark cloud is being burnt away by theta-1C. The edges of the saucer curve up and over, as can be seen in the wide-angle optical photograph (Fig. 4.13). But the bottom of the saucer is not entirely smooth. To the lower left of the Trapezium it is rucked up, and we see a portion of the glowing gas layer edge-on, so making it appear brighter.

This fortuitous alignment means astronomers can use Orion's bar to probe a cross-section of the thin sheet of glowing gas – technically, an ionisation front. **Fig. 4.17** is a detailed view of the bar in ultraviolet radiation. The region mapped extends from under the star theta-2A (bottom left) to just short of the Trapezium at top right.

The image is constructed from radiation at three specific wavelengths, emitted by carbon atoms with one electron missing (red), oxygen with one electron missing (green), and carbon with two electrons missing (blue). This is a sequence of increasing energy: it takes more energy to knock an electron out of an oxygen atom than a carbon atom, and more energy still to remove a second electron from carbon.

Within the ionisation front (upper part of Fig. 4.17), the gas is exposed to the searing radiation from theta-1C, and the carbon atoms are stripped of two electrons, giving a glow that is coded blue in this view. (The intense ultraviolet environment also strips two electrons from oxygen, which then produces the green glow seen in optical photographs.)

At the bar, the radiation runs into the dense cloud surrounding the nebula. Here, it is quickly weakened, to the point where the radiation can knock out only one electron from each atom. The combination of emissions at all three wavelengths makes the bar look white in Fig. 4.17. But it is tinged with red on the lower left, where the last vestiges of ultraviolet have penetrated into the dark cloud and are still removing single electrons from the odd carbon atom.

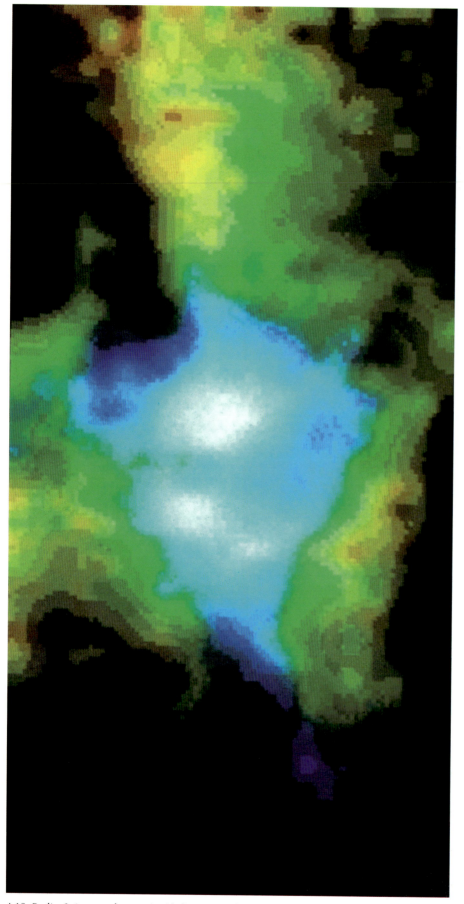

Orion infrared cluster

STARS ARE BEING born right now in the dense dark cloud of gas and dust lying behind the Orion Nebula. The dust blocks off the light from these very young stars, dimming them to less than a million-millionth of their original intensity. But the longer wavelength infrared and radio waves can cut through the dust, and reveal starbirth in action.

Molecules in the gas clouds emit radio waves, and the powerful broadcaster, carbon monoxide, is a useful guide to the extent and density of the clouds. A 150 light year long molecular cloud runs down through the lower left of Orion (Fig. 4.8), but it is only condensing at its densest part, behind the Orion Nebula. The larger scale carbon monoxide image here (**Fig. 4.18**) covers the central region of the large optical photograph (Fig. 4.13), over the same 10 light year range from top-to-bottom, but much more restricted left-to-right. It is centred just to the right of the Trapezium. The brightness of the image represents the intensity of carbon monoxide emission, and the superimposed coloration indicates the velocity of the gas – receding from us at a few kilometres per second in the blue regions near the centre, and slightly faster in the green, orange and red regions towards the edge.

The intense central region, about 3 light years across, contains enough gas to make 10 000 Suns. It is called the Orion Molecular Cloud 1 (OMC1), and radio astronomers have discovered dozens of different kinds of molecule here, some consisting of as many as seven atoms. Molecules can only survive in space when dust protects them from ultraviolet radiation, so pictures like this reveal only the dense murky dust clouds. The nebulae M42 and M43, brilliant at other wavelengths, are too hot for molecules to survive, and so do not appear. (The former, the Orion Nebula itself, would stretch diagonally across, while M43 would lie in the dark patch on the upper left.)

Fig. 4.19 is a more detailed view of the central region of Fig. 4.18, enlarged three times. This time, the emission from carbon monoxide is colour coded for brightness, from blue for the faintest parts, through red and yellow to white for the brightest.

Fig 4.19 shows gas clouds in a variety of locations, superimposed on one another. In the background is a nearly vertical strip. It is the narrow dense core of the molecular cloud seen in the previous image, running north–south behind the Orion Nebula.

The strips of emission running across the top and bottom lie in front. They wrap neatly around the hot gas of the Orion Nebula as seen in light (Fig. 4.10) and radio (Fig 4.16). The 'bar' of carbon monoxide to the lower left lies just outside the bar that we see at optical and

4.18 *Radio, 2.6 mm carbon monoxide line, 13 m telescope, Five College Radio Astronomy Observatory*

ultraviolet wavelengths (Fig. 4.17). It marks the dense gas outside the ionisation front, which forms a barrier to ultraviolet light from the star theta-1C.

Finally, the brightest spot in Fig. 4.19 shows a dense concentration of gas in the middle distance. It lies behind the optically visible nebula, just to the top right of the Trapezium. Dense dust totally hides the secrets of this region from optical telescopes, yet astronomers exploiting new astronomy techniques have discovered that this 'Orion infrared cluster' is the highlight of the whole magnificent constellation.

Fig. 4.20 shows a variety of images taken at near infrared wavelengths, which partially penetrate the dust . They give a clearer idea of how the Orion Nebula would look if all the interstellar pollution were cleared away. The images are colour coded for brightness, with dim regions green, and successively brighter parts yellow, blue, white, red and white again.

The top right image reveals hot gas at a temperature of around 10 000 K, ionised and made to glow by hot stars. Most of this radiation comes from the optically visible star theta-1C Orionis, shining on the front surface of the dark cloud, so this view is very similar to the nebula we see optically.

Small solid grains of dust at a temperature around 1000 K show up in the bottom left image. Notice that the bar of dust (at lower left) lies further out than the bar seen in the hot gas image: it coincides with the region of dense gas and dust seen at carbon monoxide wavelengths (Fig. 4.19). These two images are in fact very similar, apart from the bright spot near the middle of the 'dust' image, formed by particles lying within the Trapezium star cluster. The second bright spot, to the upper right, is the Orion infrared cluster.

The top left image shows mainly the radiation from stars in the region. The brightest object here is the Trapezium, although its individual stars cannot be distinguished in this view. Theta-2A Orionis lies to the lower left of the bar. The Orion infrared cluster shows up once more, to the upper right of the Trapezium. If our eyes were sensitive to this wavelength of radiation, the Orion Nebula would seem to have two centres, with the Orion infrared cluster as the fainter of the two. But the final image suggests its true importance.

The radiation used to make the bottom right image comes from molecules of hydrogen deep in the molecular cloud. Ordinary hydrogen molecules do not emit radiation in the infrared, but this observation was tuned to the emission from hydrogen molecules that have been energised by shock waves punching through the gas cloud. This image clearly shows that the centre of action in Orion is not the Trapezium – for all its optical glory – but the hidden Orion infrared cluster.

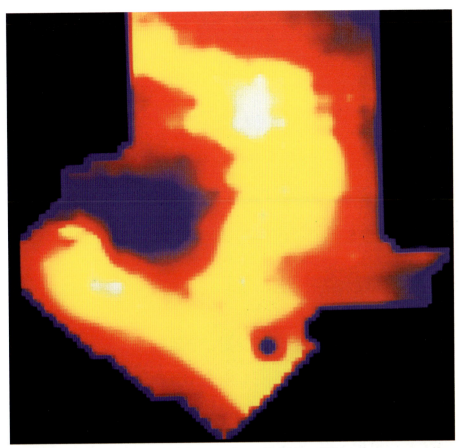

4.19 *Radio, 1.3 mm carbon monoxide line, 15 m James Clerk Maxwell Telescope*

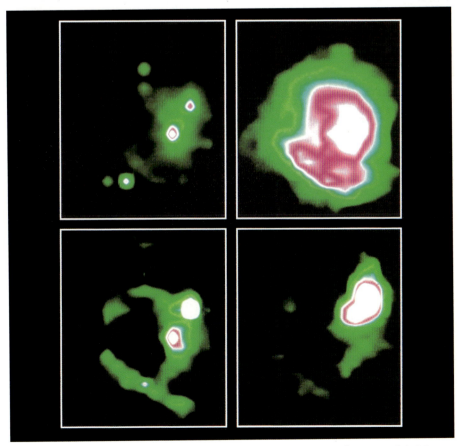

4.20 *Infrared, 2.2 μm (top left), 2.166 μm hydrogen atom Brackett gamma line (top right), 3.3 μm (bottom left), 2.122 μm hydrogen molecule line (bottom right), 3.8 m UK Infrared Telescope*

4.21 *Infrared, 1.25 µm (blue), 1.65 µm (green), 2.2 µm (red), 3.8 m UK Infrared Telescope*

Infrared radiation easily penetrates the dense dust cloud behind the Orion Nebula, revealing what lies within. New infrared cameras can show in detail the stars hidden there and a clutch of powerful newly-born stars.

Fig. 4.21 is an infrared view of the Orion Nebula, colour coded according to wavelength. It shows how the nebula would look if our eyes were sensitive to radiation with a wavelength three times longer than we actually perceive.

The bluish (short-wavelength) images in Fig. 4.21 mirror fairly well the optical view of the nebula: the central Trapezium cluster, the bright star theta-2A at the lower

left and the glowing bar next to it. But at the longest wavelength (coded red), we can see radiation that is escaping from the depths of the dark cloud behind. The more dust in the way, the redder the image appears – like streetlights dimmed and reddened by mist. So the colours in Fig. 4.21 also act as a crude coding for distance: the white images are the nearest, and the reddest images the most distant stars, deeply embedded in the cloud.

Eighty per cent of the stars in Fig. 4.21 are hidden from optical telescopes by dust. They are incredibly densely packed. Five hundred stars lie in a region only two light years across: half the distance from

the Sun to the nearest star.

The reddish region to the upper right is the Orion infrared cluster, which appeared in less detail on previous images (Figs. 4.19 and 4.20). It is most intense at longer infrared wavelengths, and was discovered at a wavelength of 22 micrometres by pioneer infrared astronomers Douglas Kleinmann and Frank Low in 1967: hence its alternative name, the Kleinmann-Low nebula, or just KL.

The finer details of this region appear in **Fig. 4.22**, which is colour coded for brightness: red parts are the most intense, and yellow and blue regions progressively dimmer. Here we are seeing a cluster of newly-born stars swathed in dust, but easily visible at a wavelength of 2.2 micrometres.

The brightest star, at the top right is the Becklin–Neugebauer object (BN), discovered by Eric Becklin and Gerry Neugebauer in 1967. BN emits most of its radiation at short infrared wavelengths (it is the brightest object in this region in Fig. 4.21), showing that it has a temperature of about 600 K. At first, astronomers thought it was a protostar – a small clump of gas and dust which is warming up as it condenses to become a star, but which does not have an energy-producing core. Astronomers have now, however, detected radiation which comes specifically from hydrogen atoms at a temperature of 10 000 K, showing that BN must contain a star which is already shining. This star is only a few thousand years old. The infrared source of BN is a surrounding ring of warm dust – about the size of the Solar System – left over from the star's formation.

The fuzzy yellow patch with a red core, below BN in Fig. 4.22, is a dusty cluster of newly-born stars. Here lies the powerhouse of Orion – but draped in such thick dust that even its infrared radiation has problems escaping.

Only at the longer infrared wavelengths do we have any hope of catching a glimpse of this lurking beast. **Fig. 4.23** is a view of KL at a wavelength six times longer. Different levels of brightness are shown by a series of colours (running from dark blue, through light blue, buff and brown, to red) repeated in succession to produce fringes like contours lines around the brightest spots.

The 'peak' at the top of Fig. 4.23 is BN once more, and another, less powerful, young star appears at the bottom. But now a new object shows up, as an irregular brown and blue peak to the left of centre. Catalogued as 'infrared compact source number 2', it goes merely by the unwieldy name IRc2.

IRc2 is the classic case of a star hiding its light under a dark bushel of dust. Detailed investigations of its spectrum show that the newly-born star IRc2 is actually ten times brighter than BN, shining as brilliantly as 100 000 Suns. It is surrounded by a tight ring of dust that absorbs practically all of its light.

This energetic young star is responsible for heating up the gas and dust in the KL nebula. It is also ejecting gas into space at a speed of several hundred kilometres per hour. Like supersonic aircraft, these clumps of gas drive shock waves through the surrounding cloud, making the hydrogen molecules shine at their characteristic infrared wavelength (Fig. 4.20). And the speeding clumps of dense gas leave streamers of glowing material in their wake, visible to the top right of Fig. 4.21 as a faint irregular 'fan' of reddish gas.

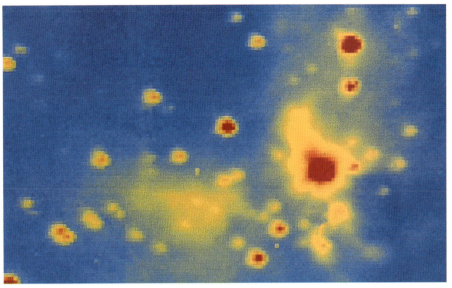

4.22 *Infrared, 2.2 μm, 3.8 m UK Infrared Telescope*

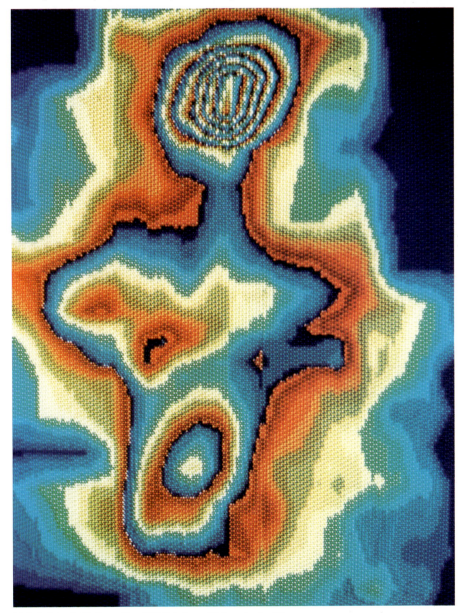

4.23 *Infrared, 12.4 μm, 3 m NASA Infrared Telescope Facility*

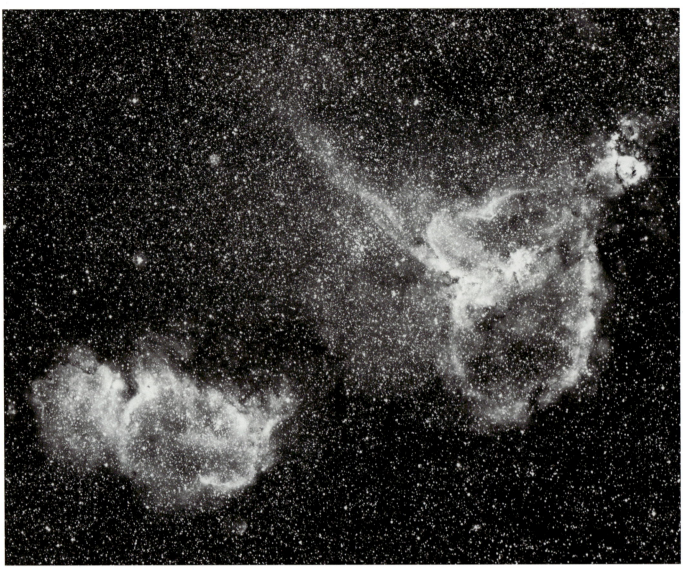

4.24 *Optical, red light, 20 cm Schmidt camera*

W3

JUST TO THE left of the well-known W-shape of the constellation Cassiopeia is a complex of faint nebulae looping around scattered star clusters. They were first picked out by the eagle-eyed American astronomer E. E. Barnard at the turn of the century, and listed in the second Index Catalogue (IC) of 1908. Yet this region is the birth place of some of the most brilliant young stars in our region of the Galaxy – stars totally obscured by dense dust clouds until revealed by the new astronomies.

Fig. 4.24 is an optical photograph of this region, taken through a filter that picks out the red glow of hydrogen in the nebulae, while subduing the light from stars. This view extends 4 degrees (eight Moon-widths) from left to right, and it shows two huge looping nebulae each over a hundred light years across. On the left is IC 1848, described by Barnard as a 'cluster with faint stars in faint nebulosity'. The bigger loop on the right, IC 1805, was a 'cluster,

coarse, extremely large, extended nebulosity'.

These nebulae are the silver lining of a huge dark cloud. It extends across the bottom right of the photograph, and can be seen by its vast silhouette against the distant stars. The cloud is thickest and densest at the right-hand edge of IC 1805. Near its northern edge (top right of Fig. 4.24) is a small triangular nebula, IC 1795, dismissed by Barnard as merely a 'patch of nebulosity'. Yet it is the signpost to the action.

The first clues came from radio astronomy. In 1958, the pioneering Dutch radio astronomer Gart Westerhout came across three large bright radio sources in Cassiopeia. Fig. 4.25 is a recent radio image of this part of the sky, colour coded so that the most intense regions are red, and fainter parts yellow, green and blue. (This image is aligned with the band of the Milky Way, so it is tipped 25° clockwise relative to Fig. 4.24.)

The strongest radio emission (red) is

evidently coming from the hot glowing gas in the looped nebulae, though the sensitive radio telescope is also picking out fainter emission from more widely-spread gas clouds all around. (Most of the small dots are distant radio galaxies and quasars far off in the background.) The numbers in the Westerhout (W) catalogue designate the left-hand loop (around IC 1848) as W5, the large central loop as W4, and the small but extremely intense region at top right, near the nebula IC 1795, as W3.

Fig. 4.26 shows the same region observed at infrared wavelengths and orientated as Fig. 4.24. The colour coding indicates temperature: red regions are coolest, at only 30 K, while the blue parts are at a couple of thousand degrees. The isolated blue dots are old giant stars.

The loops in the two big nebulae show up clearly, and at first sight the infrared view looks much like the optical photograph with the distracting stars removed. The loops are relatively cool (reddish), marking regions where dust is

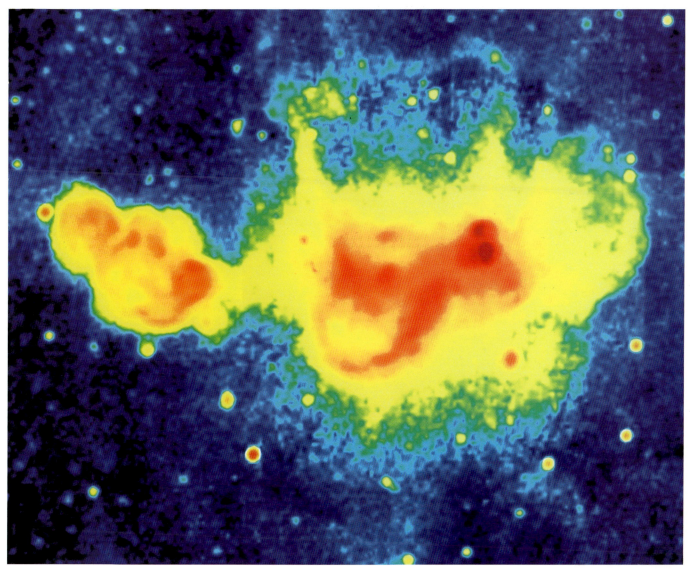

4.25 *Radio, 11 cm, 100 m Effelsberg Telescope*

mixed in the gas that glows at optical wavelengths.

But the hotter (whitish) regions of the infrared image generally correspond to *dark* patches in the optical image: note, for example, the carrot-shaped region in IC 1848 (left-hand nebula). And the right-hand rim of IC 1805 is positively brilliant: regions within the dark cloud are shining at infrared wavelengths, making this rim much wider than it appears in the optical photograph.

As in the radio view, W3 – the region near IC 1795, at top right – is by far the most brilliant. It is so bright that the detectors on the Infrared Astronomical Satellite have given it a pair of false 'spikes', like the spikes we see in photographs of extremely brilliant stars. It is a sure indicator of a massive rash of star formation, hidden in the dark clouds of W3.

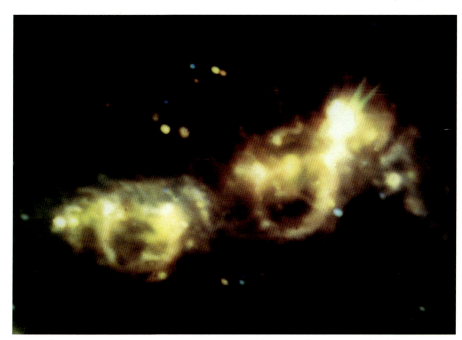

4.26 *Infrared, 12 μm (blue), 60 μm (green), 100 μm (red), Infrared Astronomical Satellite*

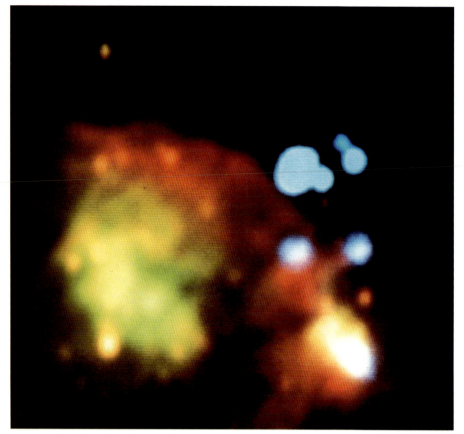

4.27 *Combined wavelengths: optical (green and red), 0.9 m reflector, Kitt Peak National Observatory; radio (blue), 20 cm, Westerbork Synthesis Radio Telescope*

4.28 *Infrared, 1.2 µm (blue), 1.65 µm (green), 2.2 µm (red), 3.8 m Mayall Telescope*

Detailed radio and infrared images can reveal the clutch of new stars being born in the W3 region of Cassiopeia, hidden from ordinary telescopes. The heart of W3 appears in **Fig. 4.27**, a composite image which combines optical with radio.

The optical view is itself a combination of two images, one showing the green light from oxygen (coded green) and the other the red light from sulphur (coded red), so giving an enhanced true-colour effect. It shows clearly the right-hand side of the triangular nebula IC 1795, magnified ten times over the wide-angle view in Fig. 4.24. The central regions of the nebula, near its hot central stars, are glowing green with oxygen light.

The blue objects in Fig. 4.27 are the brightest parts of the radio source W3. Each is a shell of hot gas surrounding a newly-born star, over 100 000 times more luminous than the Sun, but totally hidden from optical telescopes by dense clouds of dust. The biggest and brightest radio source here is W3A (with the smaller W3B abutting it).

Fig. 4.28 is an infrared image that homes right in on W3A. It appears as the large nebula at top centre, with the smaller W3B as the cluster of red dust clouds at lower right. The radiation is coming largely from dust clouds: the colour coding shows wavelength, which indicates the temperature of the dust (red being coolest) and the amount of obscuration in the densest regions (red being most obscured).

The dust cocoon of W3A appears as a reddish fringe, two light years across. At its centre is a single star, producing so much energy that at middle infrared wavelengths W3 is the third brightest object in our skies, even though it lies 5500 light years away (four times the distance to the Orion Nebula).

This massive young star is as bright as half a million Suns. Its powerful radiation – and gales of gas from its surface – are continuously enlarging its cocoon. At the top left, near the edge of the all-encompassing dark cloud, the radiation is beginning to break through. It is lighting up a wisp of gas (blue-green) outside the dense dust cloud. The hollow cavity has only to expand a little more before it breaks through, and we can see the light from the central star. In perhaps a thousand years – a short time in human history as well as astronomy – our descendants will see an extra star in Cassiopeia, extending its W-shape to a six-star zigzag.

In time, each of the other infrared sources in Fig. 4.28 will also emerge from the dark clouds of W3 as a new star. And, as the wide angle radio and infrared views (Figs. 4.25 and 4.26) suggest, there are stars forming to the south of IC 1795 as well. One particular interesting example is shown as a radio image in **Fig. 4.29**. The colour coding runs from blue for the faintest regions, through green, pink and purple to white for the most intense peak. The shape indicates the hot gas is forming a shell, only one-twentieth of a light year across. It must be heated up by a central hot young star. (Like most stars, it is not detectable at radio wavelengths).

As well as the radio emission at all wavelengths from its hot gases, this source shines exceptionally intensely at wavelengths near 18 centimetres produced by hydroxyl (OH) – a water molecule with one hydrogen atom missing – and is thus known as W3 (OH). The brightness of this emission shows that the hydroxyl molecules must be shining as natural masers, the radio equivalent of lasers.

Fig. 4.30 shows the brightest region of W3 (OH), observed at the wavelength emitted by hydroxyl molecules. The high resolution was obtained by observing with eight radio telescopes on sites from Massachusetts to California, thereby simulating an instrument the size of the United States. The entire field of view in Fig. 4.30 is only 0.15 arcseconds across, far smaller than the size of a star's image in an optical photograph of the sky.

Each source in Fig. 4.30 is a patch of gas in the shell of W3 (OH) where the hydroxyl molecules are acting together to form a maser. The colour coding shows the speed of the maser, towards us (blue) or away from us (red), calculated from the Doppler shift in the wavelength of the hydroxyl line. The blue-coded masers are on the near side of the expanding shell, and the red-coded masers on the far side. The positions, intensities and velocities of masers provide a new highly sensitive probe of starbirth regions.

The W3 region of star formation lies between the young star clusters and tattered nebulae of IC 1805 and IC 1795, on the one hand, and a large region of dark clouds that stretches an equal distance in the other direction. Here, we are seeing a progression of starbirth.

A few million years ago, this entire region was filled with diffuse interstellar gas and dust. It was then swept up by the Perseus Arm of the Galaxy, which compressed the cloud to such an extent that stars started forming at one end (the left extremity as viewed in images like Fig. 4.24). This first generation of stars heated up and dispersed the gas and dust around, and compressed the adjacent region of cloud until it, in turn, started to collapse and spawn stars.

The rash of star formation will undoubtedly continue to spread into the half of the original cloud that has survived so far (off to the right of the wide-angle images on the previous spread). Half a dozen large dense clumps in the cloud stretching away from W3 will, in time, become new clusters of stars.

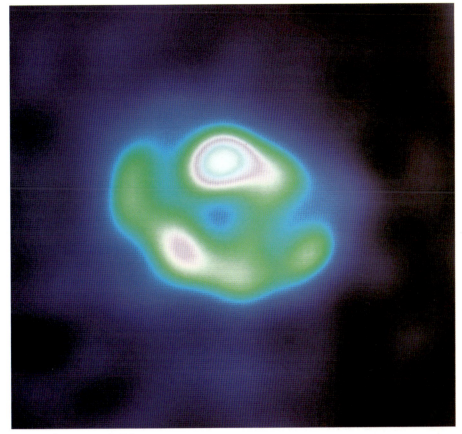

4.29 *Radio, 1 cm, Ryle Telescope*

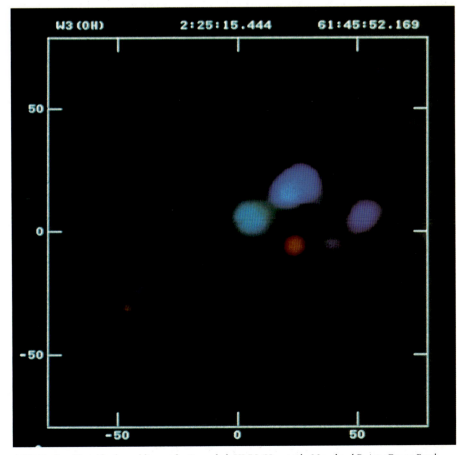

4.30 *Radio, 18 cm hydroxyl line, velocity coded, VLBI (Haystack, Maryland Point, Green Bank, Algonquin, Vermilion River, Fort Davis, Owens Valley, Hat Creek)*

4.31 *Optical, true colour, 1.2 m UK Schmidt Telescope*

4.32 *Optical, true colour, 3.9 m Anglo-Australian Telescope*

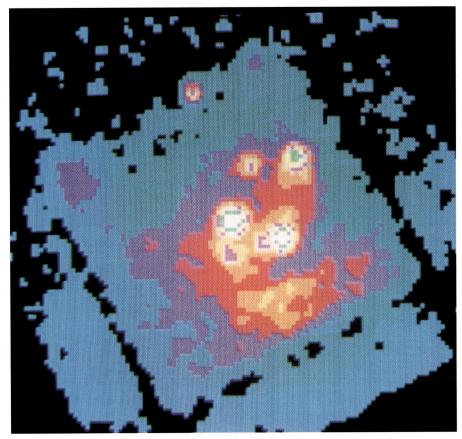

4.33 *X-ray, 0.3–2.5 nm, Imaging Proportional Counter, Einstein Observatory*

Carina Nebula

THE HUGE CARINA Nebula is visible from Earth's southern hemisphere as a misty glow in the band of the Milky Way late in the summer and autumn. The colour photograph (**Fig. 4.31**), taken by combining separate exposures through different coloured filters on the UK Schmidt Telescope in Australia, covers a region of sky about 3° by 4°. The Carina Nebula stretches over some 2 ½° (about five Moon breadths). At its distance of

9000 light years, the nebula is thus a staggering 300 light years across – twenty times the size of the Orion Nebula.

Two million years ago this nebula was a huge, dark, dense molecular cloud. Much of this matter has now condensed into stars, and the fierce radiation from the heaviest, hottest stars is lighting up the remaining gas. The tatters of the original cloud appear as the dark dust lanes which make a huge 'tick' across the nebula. In a star-formation region, stars of all masses are born, but the heaviest are very rare.

Because it is so immense, the Carina Nebula is one of the few places in our Galaxy where we find many heavy stars.

The shorter exposure photograph from the Anglo-Australian Telescope (**Fig. 4.32**) reveals the stars in the central 1° of the nebula. Many of the stars strewn over the picture lie well in front of the nebula, but there are two clusters of stars which do reside in the nebula and are largely responsible for heating it. The compact cluster (to top right of centre) is Trumpler 14, and the looser cluster in the centre Trumpler 16; both were first catalogued by Robert J. Trumpler of the Lick Observatory in the 1920s. Between them, the two clusters contain over a dozen stars heavier than twenty Suns (O type stars) – the kind of star (like W3A or theta-1C in Orion) which would be the outstanding member of an ordinary star cluster.

Amongst these awesome stars, one in each cluster contends for the title of the Galaxy's most massive and luminous star. The central star of Trumpler 14, catalogued HD 93129A, weighs at least 120 times as much as the Sun, and shines as brightly as five million Suns. Because of its high surface temperature of 52 000 K, HD 93129A emits most of its energy as ultraviolet radiation. Its rival, eta Carinae, lies in Trumpler 16, at the lower-left end of the small, banana-shaped dark cloud. It is surrounded by a shell of dust and here appears slightly oval. Eta Carinae is about as heavy and luminous as HD 93129A, but most of its radiation is emitted at infrared wavelengths.

The Einstein Observatory's X-ray view (**Fig. 4.33**) is to the same scale as the optical above. The colour coding runs from blue for faint regions, up through purple, red, orange, yellow, purple and pale blue again to white for the most intense; the dark square is due to supporting struts in the instrument. The source in the centre of the square is eta Carinae, and the X-ray emitter to the top right is HD 93129A.

The X-rays come from gas in the stars' outer atmosphere (corona), at a temperature of 10 million K, and they make up only a ten-millionth of the stars' total output of radiation. The bright X-ray source to the right of eta Carinae (HD 93162) is a star only one-tenth as massive, but going through an active phase (during which it is classified as a Wolf Rayet star). It has an exceptionally high X-ray output for a star, amounting to over a millionth of its light output. The gases from these stars' coronae are blowing a tenuous bubble of multi-million degree gas, responsible for the background 'glow' (red, purple and pale blue in Fig. 4.33) within the cooler, 10 000 K, gas of the Carina Nebula itself.

4.34 *Optical, true colour, 3.9 m Anglo-Australian Telescope*

Eta Carinae

IT IS POSSIBLY the most luminous star in the entire Galaxy, yet eta Carinae hides its light under a bushel of dust that veils its true brilliance.

Fig. 4.34 is a close-up of the central regions of the great Carina Nebula, enlarged 15 times compared with Fig. 4.32. The brightest object here (lower left) is eta Carinae, looking distinctly unstarlike: its central core is surrounded by reddish strands of gas and dust. The dark cloud to its right is the top of the banana-shaped silhouette seen in the previous images of the Carina Nebula.

Eta Carinae was relatively faint

(magnitude 4) when Edmond Halley first estimated its brightness, on a fleeting visit to the Southern Hemisphere in 1677. By the 1820s, it had brightened to second magnitude. In 1837, eta Carinae surprised the veteran astronomer John Herschel, observing from South Africa, as it put on 'the appearance of a new candidate for distinction among the very brightest stars of the first magnitude'.

Six years later, eta Carinae was almost rivalling the brightest star, Sirius, even though it lies a thousand times further away. It must have been as brilliant as five million Suns. Then eta Carinae faded to magnitude 8, and has stayed below the limit of naked eye visibility ever since.

The neighbouring nebulosity has also changed. John Herschel described the large bright-rimmed oval (upper right of Fig. 4.34) and the band of dust below it as forming a dark keyhole. The huge hot clouds of hydrogen in a nebula, however, cannot alter appreciably in the course of a mere century. Fig. 4.34 indicates what has happened. The left-hand rim of the keyhole, immediately to the right of eta Carinae, is whiter than the rest of the nebulosity. It does not consist of glowing gas, but of clouds of dust lit up by eta Carinae. In Herschel's time, the star was brighter, so the rim of the keyhole was more brilliant; as the star grew dimmer, the dusty rim faded too.

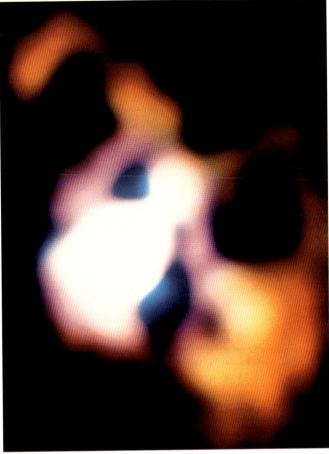

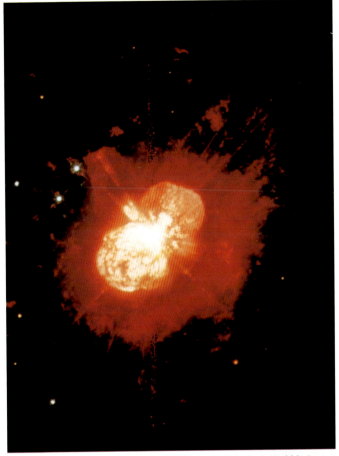

4.35 *Infrared, 2 μm (blue), 4 μm (green), 5 μm (red), COME-ON adaptive optics, 3.6 m reflector, European Southern Observatory*

4.36 *Optical, true colour, Wide Field/Planetary Camera 2, Hubble Space Telescope*

The debut of infrared astronomy in the 1960s produced a sudden resurgence of interest in eta Carinae. As seen at middle infrared wavelengths, eta Carinae is the brightest object in the sky (beyond the Solar System). Its infrared luminosity still amounts to several million times the Sun's total power output.

The infrared image, **Fig. 4.35**, shows in detail this strange star, and the tiny cocoon of dust wrapped around it. Colour coding indicates wavelength. The shorter wavelengths (blue) are coming from the star itself, either directly or reflected off dust grains which blur out the star's image. The longer wavelengths (green and red, combining to form yellow) are the heat radiation from dust grains themselves.

The bright dust cloud near the star (coded white) is known as the Homunculus – the 'little man', with his head towards the upper right. But most of the radiation is emitted from the further-flung clouds, coded orange and red, which stretch a third of a light year from eta Carinae itself. They are being heated to 250 K, almost to the melting point of ice and a hundred times hotter than the background temperature of space.

Both the Homunculus and debris scattered further afield are apparent in a view from the Hubble Space Telescope (**Fig. 4.36**). In this true-colour image, the dusty central regions appear white, while the outer regions glow red with emission from nitrogen atoms. Long thin jets stand out clearly in Fig. 4.36, showing where the star has ejected small 'bullets' of gas and dust.

Wrestling with blurred ground-based views, astronomers have argued for years about the geometry of this cloud. Is it a double outflow of dust, like the double jets seen in many astronomical objects? Or is it a single thin disc, viewed at an angle?

The Hubble view supports an interpretation by British astronomer David Allen: the dust forms an hourglass shape (or double peanut). The hourglass is tipped so its lower lobe points towards us, while the upper lobe lies behind the central star. Allen found these lobes are expanding at 660 kilometres per second. The eruption was presumably confined to this shape by a dense disc of gas and dust (not seen here) that circles the star around the waist of the hourglass.

Optical astronomers have patiently tracked some of the brighter regions in the Homunculus over several decades, and have concluded that the small dusty nebula was ejected during the star's most brilliant period, around 1840.

Astronomers can explain all these tantalising clues by a single scenario. Eta Carinae is among the most massive stars in the Galaxy, weighing as much as a hundred Suns. Its surface temperature is 29 000 K, and it produces as much energy as five million Suns, mainly in the form of light and ultraviolet radiation. This was the star as seen directly in the 1820s. But around 1840, its outer layers become unstable, and blasted off into space. Gaseous elements quickly cooled and condensed as dust grains. The central star survived the eruption, but the blanket of dust hid the central star almost completely from sight. In doing so, the dust grains heated up, and they now reradiate the central star's luminosity as infrared.

The final piece in the puzzle concerns the gas around eta Carinae. It consists largely of nitrogen, an element that is only abundant well inside a massive star: it seems that eta Carinae has already undergone several eruptions that have stripped away its outer layers.

A heavyweight like eta Carinae rips through its nuclear fuel at a prodigious rate. It is now nearing the end of its life, even though it lies in a 'star nursery' where portions of the Carina Nebula are still condensing into stars. Some time in the comparatively near future, eta Carinae will have one final moment of glory as it self-destructs in a supernova explosion.

4.37 *Optical, true colour, 1.2 m UK Schmidt Telescope*

The Pleiades

'MANY A NIGHT I saw the Pleiads, rising thro' the mellow shade,
Glitter like a swarm of fireflies, tangled in a silver braid.'

The words of Alfred, Lord Tennyson, in the poem Locksley Hall, provide the most eloquent description of this glorious little star cluster. But it has excited interest since the earliest times. The Chinese recorded the Pleiades in 2357 BC, and it is one of the few astronomical names to occur in the Bible, when God asks Job 'canst thou bind the sweet influences of the Pleiades?'

Since ancient Greek times, the cluster has been dubbed the Seven Sisters. The true colour photograph **Fig. 4.37** reveals many more 'sisters' than that! The cluster contains at least 300 individual stars. The brightest are hot blue-white giants, while the fainter members are cooler and more yellow. From this distribution of star colours, astronomers have deduced that the cluster is some 70 million years old.

And that raises the 'Pleiades paradox'. Although the cluster is young compared to the Sun's venerable 4600 million years, it is old enough to have moved well away from the nebula of its birth. Yet the cluster is

immersed in glowing nebulosity. Moreover, as we can see in Fig. 4.37, these patches are not shining red or green, like the central regions of the Orion Nebula (Fig. 4.13), but a distinct blue. And they are combed into strands running for several light years.

The answer to the paradox appears at infrared wavelengths. **Fig. 4.38** is a view from the Infrared Astronomical Satellite (IRAS), colour coded for wavelength, which is also an indicator of temperature: blue represents emission from gas and dust at around 250 K, while red emission arises in interstellar material only 30 degrees above absolute zero. Fig. 4.38 covers a region 6° square (12 Moon-widths), about four times wider than Fig. 4.37, and the Pleiades cluster sits entirely within the bright white area of the infrared image.

After its birth, the Pleiades shrugged off its original nebula and lay – like other clusters – in empty space. But a few million years ago, a huge interstellar cloud laced with magnetic field bore down on the Pleiades and enveloped the cluster. The cold dust grains in the cloud shine red in the colour-coded infrared view, Fig. 4.38, except in the region of the Pleiades, where they are warmed by the stars' radiation and glow a brilliant white in the IRAS image.

As the huge cloud has moved past the Pleiades, travelling from right to left in Fig. 4.38, the star cluster has swept clear a long 'wake'. It extends to the left in the IRAS view, as an elongated black region devoid of gas and dust, like the downstream wake from an ocean liner.

The nebulosity seen in the optical image, Fig. 4.37, is thus not associated with the Pleiades at all. It just happens to be passing by. The Pleiades stars are illuminating streamers of dust arranged in hair-like strands by the cloud's magnetic field. The nebulosity appears strikingly blue because the small particles of dust scatter blue light more effectively than red: called Rayleigh scattering, it is similar to the mechanism that gives our sky its blue colour.

The tenuous gas in the cloud does not shine noticeably, except at the bottom where a slightly denser portion is heated by Merope, the southernmost bright star in the Pleiades. Merope is the hottest star in the group, at 15 000 K, though not the biggest and brightest. That honour belongs to the star to the left of Merope, Alcyone.

Many of the stars in the Pleiades have very hot coronae, or outer atmospheres, which show up in the X-ray image (**Fig. 4.39**). Here, the X-ray brightness is

4.38 *Infrared, 12 μm (blue), 100 μm (red), Infrared Astronomical Satellite*

colour coded so that faint stars are blue, and the centres of successively brighter stars reach green, yellow, red and white. This view covers just the central half a degree of the Pleiades: the only two bright stars that would appear here are Merope near the bottom right and Alcyone at the middle left.

In fact, neither of these stars stands out at X-ray wavelengths. The most prominent X-ray sources correspond with some of the faintest stars seen optically, and particularly with those that rotate rapidly. The brightness of the corona, it turns out, does not depend much on the star's output of light from its photosphere, but does depend on its spin rate.

The connection, astronomers believe, is magnetism. A rapidly spinning star winds up and amplifies its internal magnetic field, which breaks through the photosphere in powerful loops that dump energy into the corona and heat it up. As stars grow older, they spin more slowly and the coronae become less powerful X-ray sources. The corona of our middle-aged Sun can still sometimes disrupt life on Earth with a powerful flare or storm: conditions on a planet circling one of the Pleiades must currently be lethal!

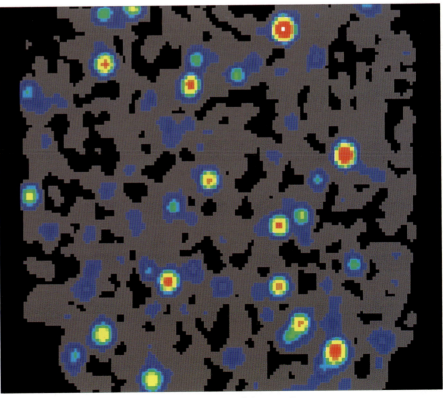

4.39 *X-ray, 0.6–10 nm, Position Sensitive Proportional Counter, Rosat*

5 infrared astronomy

IN 1800, SIR William Herschel – best known for finding the planet Uranus – turned his attention to the amount of 'heat' carried by the different colours in sunlight. He spread sunlight out into a rainbow spectrum, and measured the temperature rise when a thermometer was placed in each colour in turn. Herschel also moved the thermometer beyond the red end of the spectrum – and found that here too the thermometer was receiving heat, carried by 'invisible' radiation. It was the first evidence for radiation other than light, and Herschel named the invisible rays 'infrared'.

The infrared region of the spectrum covers a wide range of wavelengths (Fig. 5.1). The broadest definition encompasses waves from the border with red light (about 700 nanometres) up to one millimetre, where radio waves begin. The most convenient unit for infrared is the micrometre (one thousandth of a millimetre), colloquially called a *micron*, and the traditional limits for the infrared are then 0.7 to 1000 microns .

Astronomers – for reasons discussed below – usually restrict the term 'infrared' to a more limited range of wavelengths, from 1.1 to 350 microns. Even so, this is over a hundred times greater than the wavelength range for visible light, and over the infrared band the appearance of the sky changes completely.

Although the Sun produces so much infrared that we can

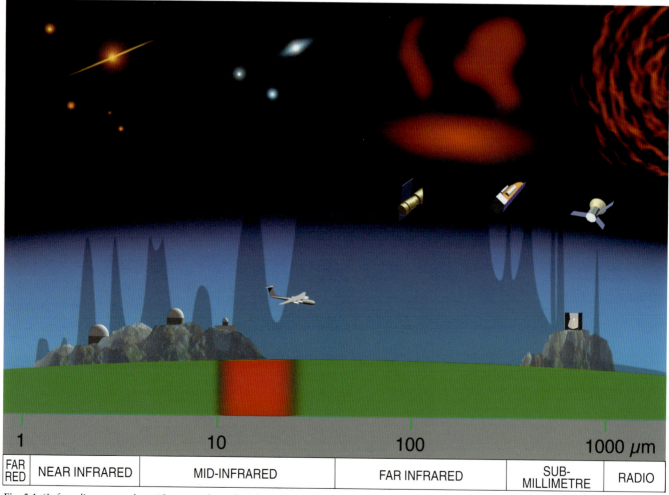

Fig. 5.1 *'Infrared' covers such a wide range of wavelengths – from around 1 to 1000 micrometres (µm) – that astronomers divide it into different regions. Far red is really an extension of the optical spectrum. All the rest of the infrared spectrum is absorbed as it comes down through the Earth's atmosphere. At the near infrared wavelengths emitted by newly-born stars, some clear 'windows' are accessible to telescopes on high mountains, such as 4200 m Mauna Kea in Hawaii. In the mid-infrared, astronomers must cope with the glow of heat radiation emitted by their surroundings and the telescope itself. NASA's airborne observatories can fly above the absorption and view, for example, distant starburst galaxies. The far infrared is totally unobservable from the ground, and astronomer s must rely on satellites: the Infrared Astronomical Satellite (left) and the Infrared Space Observatory (middle) have investigated cool interstellar clouds about to condense into stars. The Cosmic Background Explorer (right) picks up the even colder radiation from the Big Bang. These submillimetre waves can penetrate down to mountain tops and border the radio window through the atmosphere.*

Fig. 5.2 *When seen at wavelengths a hundred times longer than visible light, the central region of the Milky Way takes on an entirely new look. In this view from the pioneering Infrared Astronomical Satellite, the radiation is cutting through the obscuring dust in space and revealing cool stars (blue dots) and dust clouds throughout the entire depth of the Galaxy. The large clouds are nebulae comparatively near the Sun: the Lagoon Nebula (lower left) is the only object that appears in an optical photograph of the same region (Fig. 8.11). The centre of the Galaxy, dimmed to invisibility at optical wavelengths, appears here in its true brilliance, with streamers of dust suggesting violent activity.*

feel its heat on our own skins, and it affects a simple thermometer like Herschel's, its strength is only due to the fact that the Sun is so near to us on the cosmic scale. Very sensitive detectors are needed to pick up the infrared radiation from any other star or from the planets, nebulae and galaxies. The only exception is the Moon, whose infrared emission was detected in the mid-nineteenth century – again because the Moon is so close to the Earth. Otherwise, infrared astronomy languished for a century and a half after Herschel's discovery until physicists began producing sufficiently sensitive infrared detectors in the 1950s. This new branch of astronomy has made tremendous strides in the past forty years, revealing aspects of the universe hidden to detectors of other wavelengths.

The easiest infrared rays to detect and measure are those that lie just beyond the red end of the visible spectrum, with wavelengths between 0.7 and 1.1 microns. Although we cannot see these rays, astronomers can image the sky at these wavelengths with special photographic plates and with the CCD detectors that have now almost completely taken over in optical astronomy. Astronomers now regard these rays as an extension of the optical spectrum, and call them 'far red'. Infrared astronomy proper starts at wavelengths longer than 1.1 microns.

As seen in the far red, the sky does not look very different from the sky seen by the human eye. It is still studded with stars, although when viewed in this new waveband their brightness is altered. Bluish stars, like Rigel in Orion, are dimmer, while red stars shine more brightly. Rigel's neighbour in Orion, the intensely red Betelgeuse, becomes the brightest star in the sky, outshining at wavelengths of one to two microns all other stars, including the visually brightest star Sirius.

When we observe other galaxies in the far red, there is little sign of the compact clumps where newborn stars blaze blue-hot. Instead, we see how the dim red stars, making up the bulk of its stellar content, are distributed. They usually trace out much broader and smoother spiral arms.

The borderland far-red region does, however, share one important characteristic with the true infrared. These rays can penetrate the dust in space that obscures our view of distant stars, and of stars embedded within dense clouds of gas and dust. The dust particles are about the same size as the wavelength of visible light, and they block it most efficiently. Even in the nearly empty space between the stars, dust particles dim a beam of light to less than half its original brightness over a distance of three thousand light years, a small fraction of our Galaxy's 100 000 light year extent. Dense clouds of dust can dim light so much that any stars within are totally invisible to our eyes and to traditional photographic plates. But light at far red wavelengths can penetrate this dust quite readily. CCD cameras are now photographing stars in regions of space which had previously been hidden from our view.

As we move into longer wavelengths – into the astronomers' infrared band – the sky changes dramatically. The familiar light-emitting stars fade, and we begin to detect new objects. Infrared also begins to lose its popular connotation as 'heat radiation': although a hot fire emits radiation of a few microns wavelengths, observations at progressively longer wavelengths reveal objects that are cooler and cooler, to the point of being downright cold!

From 1.1 to 4 microns, we are in the *near infrared* region. If we could view the Universe with near-infrared eyes, we would see objects with a temperature of 1000 to 2000°C. These include both the largest and the smallest of normal stars: at one end of the range the coolest of the red giant stars, and at the other end the lightweight red dwarfs with feeble nuclear reactors producing the dimmest of glows. Astronomers are also scanning the skies at these

Fig. 5.6 *The Infrared Astronomical Satellite (IRAS) – seen here undergoing prelaunch tests – has a 0.6-metre reflecting telescope. The mirror is to the left (hidden within the satellite); the open end (right) is protected here by a convex cover, ejected once in orbit. The large cowl shades the telescope from the Sun's infrared; the bulky middle section is a tank of liquid helium for cooling the telescope and detectors; the satellite's controls are in the short cylindrical section (far left).*

Infrared Catalogue (IRC), colloquially known as the Two Micron Survey. It contained 5612 cool stars – hardly more than you can make out at optical wavelengths with your unaided eye.

A survey of the sky at mid or far infrared wavelengths needed a telescope flown above most of the atmosphere. Fortunately for astronomers, the US Air Force in the 1970s had its own reasons for wanting to know the distribution of natural infrared sources: it would be rather embarrassing to confuse the heat radiation from a descending ballistic missile with the infrared emission of a distant starbirth region. In the course of ten rocket flights, the Air Force survey picked out 2000 infrared sources at wavelengths between 4 and 27 microns.

But the best way to survey the sky at the longer infrared wavelengths is from an Earth-orbiting satellite. For this purpose, infrared astronomers from the United States, the Netherlands and the United Kingdom constructed the Infrared Astronomical Satellite (IRAS) which, during 1983, studied the sky from its orbit at an altitude of 900 kilometres. But even IRAS could not escape the problem of having a 'luminous telescope'. The satellite was basking in sunlight, and at some wavelengths its own heat radiation would have swamped the feeble infrared radiation from space. The designers had to enclose the 0.6-metre diameter IRAS telescope (Fig. 5.6) in a large jacket filled with 70 kg of

liquid helium at a temperature of only 16 K, and with the detectors themselves cooled to 2 K. The mission ended after ten months, when all the helium had boiled away.

Although astronomers had flown satellites to measure other radiations from space (X-rays, for example), IRAS was one of the most difficult missions ever attempted because of the difficulties of keeping its huge amount of helium coolant – about a human's weight of liquid – near the absolute zero of temperature, out of reach of any intervention. But the returns repaid the effort handsomely. IRAS surveyed the sky at a wide range of infrared wavelengths from 8 to 120 microns – the mid and far infrared wavelengths, which are difficult or impossible to study from the ground – and discovered half a million objects emitting infrared radiation. A large proportion were previously completely unknown.

The IRAS discoveries encompassed the whole of the known Universe. The satellite picked out a comet that passed very near the Earth in 1983, and a strange asteroid – Phaethon – that skims the Sun and leaves meteor particles in its wake. IRAS pinpointed thousands of young stars and objects that may be protostars. To astronomers' surprise, IRAS also picked out infrared radiation coming from the vicinity of some mature stars, such as Vega and beta Pictoris (Fig. 5.7). These turned out to be discs of gas and dust, similar to the 'solar nebula' from which the Sun's

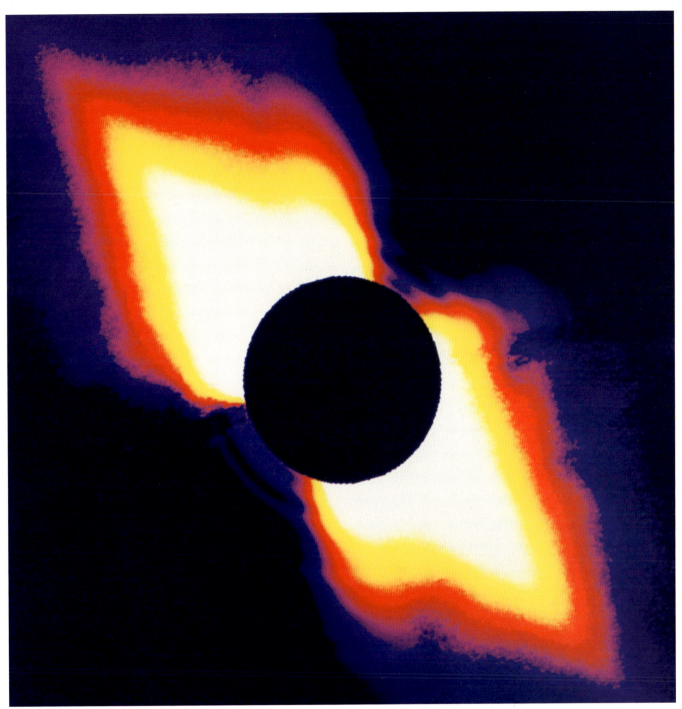

Fig. 5.7 *A strange infrared glow around the star beta Pictoris provided the first evidence of a planetary system accompanying another star. The Infrared Astronomical Satellite detected this unexpected emission in 1983: ground-based telescopes later revealed it comes from two faint 'wings' on either side of the bright star, which has here been hidden behind an opaque circular disc. These wings are an edgewise view of a disc of dust orbiting beta Pictoris. The dust is probably condensing into planets, just as dust around the Sun condensed into the worlds of the Solar System some 5 billion years ago.*

family of planets was born. The IRAS discovery suggests that at least half of stars like the Sun have planets.

Beyond the confines of our own Milky Way, IRAS picked out thousands of other galaxies. It showed where stars are being born in nearby galaxies like Andromeda. But its greatest extragalactic achievement was in finding a whole new kind of galaxy. A *starburst galaxy* is in the throes of an immense burst of star-formation: up to half the galaxy's gas may be turning into stars in one go, releasing an enormous amount of radiation mainly at infrared wavelengths. While the Milky Way – as a whole – generates about as much

infrared as visible radiation, the most extreme starburst galaxies produce a hundred times as much 'heat' as 'light'.

IRAS was designed to pick out individual infrared sources: as it scanned the sky, each source would make a 'blip' in its detectors. But the satellite's detectors turned out to be so stable that astronomers could accurately compare the infrared brightness of parts of the sky some distance apart. This gave an unexpected bonus: successive scans could be combined in a computer to produce wide-angle views of the infrared sky.

After IRAS, infrared astronomy came of age. The late

1980s and early l990s saw a blossoming of infrared astronomy, like the explosion of optical astronomy in the time of Galileo and radio astronomy in the 1950s and 1960s. It owed much to IRAS, which showed infrared astronomers where to look to find exciting new results. But it was also due to the development of the first detectors that could show what infrared sources really looked like.

With its longer wavelength, infrared radiation carries less energy than light. Astronomers need a material sensitive to very small changes in energy – and cooled to very low temperatures – to detect infrared at all.

The answer lies in semiconductors. These substances have an enormously larger resistance to electricity than metals, but are not complete insulators. More important to astronomers, radiation from outside can affect the way that electrons move about inside a semiconductor.

Semiconductors soon supplanted the lead sulphide used for the Two Micron Survey with more sensitive materials bearing exotic names such as indium antimonide, gallium arsenide and mercury-cadmium telluride (known as 'mer-cad' for short). For the longer infrared wavelengths, astronomers use crystals containing combinations of silicon or germanium, 'doped' with small amounts of other elements.

Until the mid-1980s, the detector on an infrared telescope consisted of only a single small chip of semiconductor, which could measure the infrared brightness at just a single point in the sky. This was fine for measuring the brightness of a red giant star, say, at different wavelengths. But an astronomer wanting an image of a warm dust cloud had to patiently scan the telescope back and forth across the source, and build up the picture bit by bit. Astronomers at the Anglo-Australian Observatory (see Chapter 3) and at the University of Wyoming perfected this slow laborious method, and produced the first good infrared views of the Universe.

Meanwhile, the American military was developing infrared cameras for use after dark on battlefields. As the detectors in these devices gradually became declassified, astronomers leapt at the chance to illuminate the even blacker scenes in the infrared Universe.

All these infrared detectors work rather like the CCD chip used in optical astronomy (see Chapter 3). They are divided into thousands of individual *pixels*, each measuring the brightness of the radiation falling on it. Electronic circuits built into the chip 'read' each pixel in turn, and inform a computer which can then recreate the image in the form of pixels on a monitor screen.

While optical astronomers now use just one basic design of CCD, infrared detectors come in a bewildering variety.

Fig. 5.8 *This cluster of electronics and cooling systems behind the mirror of the UK Infrared Telescope comprises the pioneering infrared camera IRCAM, which captured the first images at near infrared wavelengths. Here, IRCAM's designer, Ian McLean, adds liquid helium to keep its 'retina' – a semiconductor chip – just a few degrees above absolute zero.*

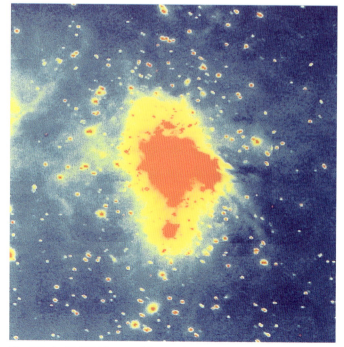

Fig. 5.9 *An IRCAM image of the star-forming region Monoceros R2, which lies near Orion, at a wavelength of 2.2 micrometres. The colour coding here represents brightness, with the most intense regions red, fainter parts yellow and the background sky blue. Hidden from optical telescopes in a dark cocoon, this nest of young stars and its surrounding wisps of warm dust are revealed in detail by IRCAM.*

1995 (Fig. 5.10). It has a mirror the same size as IRAS, and a similar cooling system. The difference comes in the sophisticated detectors: two cameras and two spectrographs will between them take detailed images and analyse radiation over a range in wavelength from 3 to 200 microns.

The American Space Infrared Telescope Facility has been under discussion – and delay – since 1971. It is intended as the last of NASA's four Great Observatories, after the Hubble Space Telescope and the Compton Gamma Ray Observatory (both now in orbit) and the planned Advanced X-ray Astrophysics Facility. The plans for SIRTF have been whittled down over the years, but with the great improvements in infrared detectors it should be able to keep its original promise to astronomers.

Even so, both ISO and SIRTF will only have a lifetime of two to three years before their cooling helium runs out. Researchers are already working on the next phase of infrared astronomy: a satellite that uses solar power to keep cool – rather like an orbiting refrigerator. Then will be the time for a large and long-lived infrared observatory in space that will bring this particular new astronomy to full maturity.

Different observatories have chosen types that match their research needs. Astronomers at America's Kitt Peak National Observatory have pioneered the platinum silicide array in astronomy. This contains a huge number of pixels, and so shows a comparatively large area of sky in detail, but it responds to only one infrared photon in a hundred. The device has thus produced stunning views of some of the larger and brighter objects, such as nebulae and the galactic centre.

More sensitive, but with a smaller field of view, is the mer-cad chip at the heart of the IRIS detector on the Anglo-Australian Telescope. It has captured exquisite shots of distant nebulae in our Galaxy and of galaxies beyond.

The third of the new detectors for the near infrared – and the earliest to be up and running – was IRCAM (Fig. 5.8), the infrared camera installed on UKIRT in 1986. It has a smaller field of view again, but is highly efficient in detecting photons – with a hit rate of two in three – and so is ideally suited to finding the faintest sources.

The first array working in the difficult mid-infrared was actually operating even earlier, thanks to the persistence of a group at NASA's Goddard Research Center in persuading the Department of Defense to hand over a suitable detector. And with each passing year of the infrared astronomy boom, better and larger arrays for both mid and far infrared astronomy are becoming available.

For astronomers, one driving force has been the promise of observatories in space. These will have all the benefits that IRAS enjoyed, and will build on its huge catalogue of discoveries by investigating infrared sources in detail – including a thorough study of their spectra at infrared wavelengths.

The immediate successor to IRAS is the European Infrared Space Observatory (ISO), launched in November

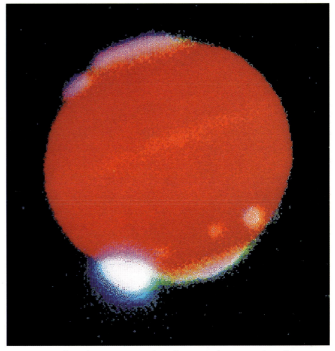

Fig. 5.11 *Infrared astronomy came of age in July 1994, when the fragments of Comet Shoemaker-Levy 9 crashed into Jupiter. The impacts threw up huge plumes that were faint to optical telescopes but outshone Jupiter itself when viewed at infrared wavelengths. This image was obtained at the Mount Stromlo Observatory in Australia, and is at wavelengths ten times longer than the eye can see (3.4-4.0 micrometres). It shows the plume from impact K rising 3000 kilometres above the clouds of Jupiter and spreading sideways to cover a region larger than the Earth.*

6 stardeath

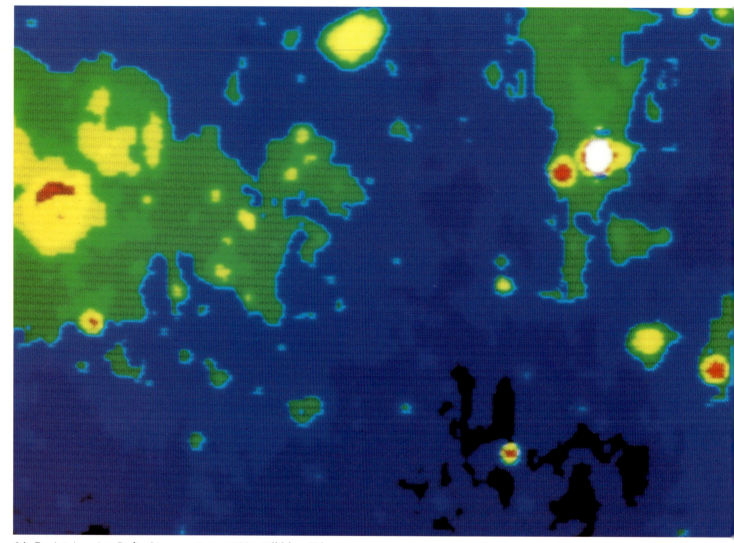

6.1 *Cassiopeia region. Radio, 21 cm continuum, 100 m Effelsberg Telescope*

THE MANNER OF a star's death depends on its mass. A relatively lightweight star like the Sun dies quite quietly. After it has expanded to become a red giant, it puffs off its gaseous envelope, as a huge, slowly expanding shell. The dense core is left as a tiny white dwarf star, with a temperature of around 100 000 K, emitting most of its radiation as ultraviolet. This heats up the surrounding shell, making its gases emit light and radio waves, while the rather cooler dust in the shell shines brightly in the infrared. Through a small optical telescope, such round shells look rather like the planet Uranus or Neptune, and they are called planetary nebulae.

A star more than six times heavier than

the Sun has a more eventful life and a spectacular death. As a red giant star, its core becomes hot enough to convert helium into a multitude of heavier elements. Within the helium core, nuclei combine to make carbon and oxygen; later, the hottest central regions convert carbon and oxygen nuclei to heavier elements like silicon; and the central temperatures and pressures within the silicon core eventually convert these nuclei to iron. At this point, the star's structure resembles an onion, with concentric shells of different composition from the external unchanged layer consisting mainly of hydrogen, through interior shells of helium, carbon and oxygen, and silicon, to the iron core.

At each stage in the giant star's evolution, nuclear fusion reactions in the very centre have provided energy to support the core against the pressures generated by the weight of the star's overlying layers. But iron has the most stable of all nuclei: iron nuclei will not take part in further nuclear fusion reactions. The upshot is dramatic. The central iron core collapses into a ball of neutrons only 20 kilometres across. This neutron star is so dense that a pinhead of its material would weigh a million tonnes.

The energy from this central collapse heads outwards, hurling the star's outer layers into space at a tremendous rate – around 7000 kilometres a second. In this

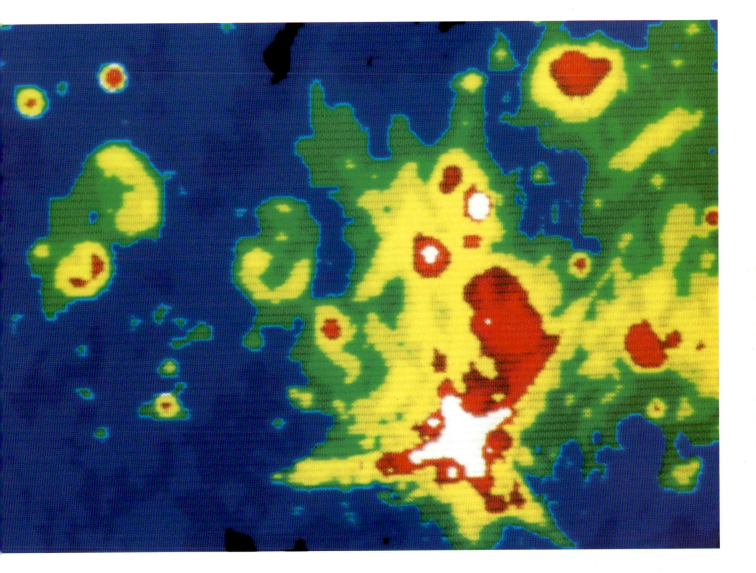

colossal explosion, a type II supernova, the star brightens up to become temporarily equivalent to 500 million Suns.

Type I supernovae are even more brilliant, and they throw out their gases at 11 000 kilometres a second. A type Ia supernova is an old white dwarf, drawing gases from a companion star until it becomes unstable and starts to collapse uncontrollably. Its contents of carbon and oxygen fuse directly into iron, in a reaction so rapid that the collapse is halted, and the white dwarf is blown apart. A type Ib is the exploding core of a red giant that has been stripped of its outer layers.

The high-speed gases from a supernova sweep up the tenuous interstellar gas, and form huge shells – the supernova remnants. These shells of gas are too hot to emit much light, but they shine brightly at X-ray and radio wavelengths. Young remnants appear as small, intense rings to X-ray and radio telescopes; older ones are larger and fainter.

The radio view of part of the Milky Way which lies in Cassiopeia (**Fig. 6.1**) is dominated by supernova remnants. It covers 12° from left to right (24 Moon breadths), with the Milky Way running horizontally. It was made at a wavelength of 21 centimetres (but not at the precise wavelength of the hydrogen line at 21.106 centimetres). Black represents the dark background and blue the general Milky Way emission, with more intense regions green, yellow, red and white. On the lower right is Cassiopeia A (Figs. 6.22–6.26), the strongest radio source in the sky (the cross shape is an artefact of the telescope). It is the remains of a supernova of the seventeenth century. Tycho's supernova remnant (Figs. 6.19–6.21), seen as a white spot above the centre, is a century older.

Very ancient remnants have expanded more and are visible as 'rings' on this scale: for example, the remnant just to the right of centre, CTB 1, is over 10 000 years old and a degree across. The red semicircle on the extreme right is the strange remnant G109.1–1.0, which contains an active neutron star (Figs. 6.39–6.41).

6.2 *Optical (5 March 1987), true colour, 1.2 m UK Schmidt Telescope*

Supernova 1987A

ON 23 FEBRUARY 1987, a four-century vigil by astronomers came to an end. A star exploded right on our cosmic doorstep, as a supernova bright enough to be seen by the unaided eye (**Fig. 6.2**). But more important, it was close enough to be accessible to all the instruments of the new astronomy. The last naked eye supernova was seen in 1604, a few years before Galileo turned his telescope to the sky.

Despite the high-tech follow-up, this supernova was discovered by nothing more sophisticated than the human eyeball and the photographic plate. At the Las Campañas Observatory in Chile, Canadian astronomer Ian Shelton found the supernova as an unusual star on a photographic plate of the Large Magellanic Cloud, the Milky Way's neighbour galaxy.

Fig. 6.2 shows Supernova 1987A (lower right), in its home environment. This small region of the Large Magellanic Cloud (see Fig. 8.19) is dominated by the Tarantula Nebula (upper left). In photographs, the supernova's small image is burnt out: in reality, it appeared far brighter than the Tarantula Nebula (Figs. 8.24–8.27).

The first signals from the supernova had actually reached the Earth several hours earlier. They were detected in two huge underground tanks of water, in Ohio and in Japan. At 41 seconds past 7.35 am (UT) on 23 February 1987, a fusillade of flashes began to fire in both tanks. Over the next 12 seconds, the Japanese Kamiokande instrument recorded 12 flashes, while eight were observed by the American Irvine-Michigan-Brookhaven (IMB) tank. **Fig. 6.3** reconstructs one flash in the IMB instrument. The frame represents the outline of the tank and the yellow dashes the region where the flash occurred. It was detected by hundreds of photocells looking into the tank.

The flashes were caused by a flood of penetrating neutrinos, which hardly ever interact with matter. For the IMB detector to stop eight neutrinos, a total of 300 million million neutrinos must have passed through the tank.

Their energies revealed that these neutrinos were born in an inferno with a temperature of 50 000 million degrees. They provided the first direct view of the central core of a star collapsing to become a neutron star. For the few seconds it lasted, the flood of neutrinos from Supernova 1987A carried away almost as much energy as the combined light of all the stars in the observable Universe.

The energy from the collapsing core ripped through the star in a matter of hours, blowing its outer layers into space. Astronomers swung all the world's major telescopes and astronomy satellites into action to observe the supernova.

Observing from Chile a month after the explosion, Peter Nisenson used speckle interferometry to sharpen his view of the supernova. This technique beats the blurring effect of Earth's atmosphere by taking thousands of very short exposures, and combining them in a computer. To his amazement, the final image (**Fig. 6.4**) showed two objects. Colour coding shows the brightest region pale yellow, and darker regions yellow, blue, green and dark blue.

The lower object, the Mystery Spot, was one-tenth as bright as the supernova itself. Far brighter than any star in the Large Magellanic Cloud, the Mystery Spot must have been related to the supernova. Possibly it was a cloud of dust 17 light days from Supernova 1987A, lit up by a beam of high-speed particles from the supernova.

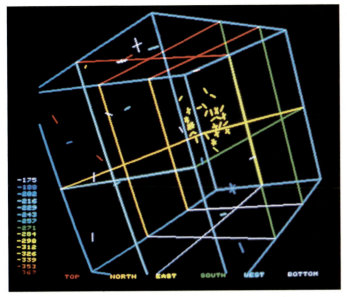

6.3 *Neutrino detection (23 February 1987), IMB detector, Ohio*

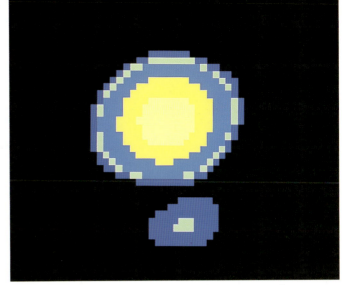

6.4 *Optical (25 March 1987), speckle interferometry, 4 m reflector, Cerro Tololo Interamerican Observatory*

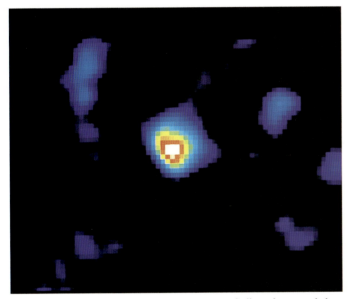

6.5 *Gamma ray (12 April 1988), 0.001–0.05 nm, balloon-borne coded mask telescope*

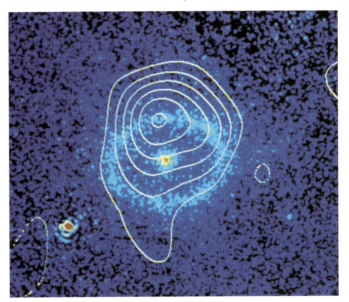

6.6 *Radio (25 June and 6 July 1991), 3.5 cm, Australia Telescope*

After the explosion in February, Supernova 1987A faded slightly – and then, to astronomers' surprise, began to brighten again. It reached its maximum brilliance in May 1987, when it shone as brilliantly as 250 million Suns.

Some new source of energy was clearly lighting up the gases from the explosion. The favourite theory invoked radioactivity. As the shock wave from the explosion ripped through the star, it could weld together nuclei of oxygen to form radioactive nickel-56. Carried out in the shell of exploding gases, these unstable nuclei would break down to cobalt-56, which in turn would decay to stable iron-56. The energy expected from these radioactive decays fitted well the supernova's changing brightness.

Final proof came late in 1987. A gamma-ray telescope, carried by balloon above much of the Earth's absorbing atmosphere, detected gamma rays from the supernova. Fig. 6.5 is an image from a flight a few months later: faint regions are blue, and successively brighter parts yellow, red and white. These images – some of the first ever made at gamma-ray wavelengths – showed that the radiation was definitely coming from the supernova rather than the X-ray source LMC X-1, which happens to lie nearby. The spectrum of the gamma rays also fitted precisely with predictions from the radioactivity theory, clearly showing emission from nuclei of cobalt-56.

As the radioactivity has decayed and the expanding gases cooled, the supernova has gradually faded from the view of gamma-ray and optical telescopes alike. But it is due for a renaissance at radio wavelengths.

Astronomers in Australia are now monitoring an ever-increasing radio flux from the supernova.

The contour lines in Fig. 6.6 show the radio intensity coming from the region of the supernova in 1991, superimposed on an optical image. The peak of radio emission lies well to the north of Supernova 1987A itself, the 'star' inside the elliptical ring of matter (Fig. 6.7).

The radio waves are generated as gases from the supernova crash into the surrounding interstellar gas. The energy of motion is converted into magnetic field and speedy electrons, which generate copious radiation by the synchrotron process. We are witnessing the birth of a supernova remnant which will, over the next few centuries, grow to resemble the powerful radio source Cassiopeia A (Fig. 6.23).

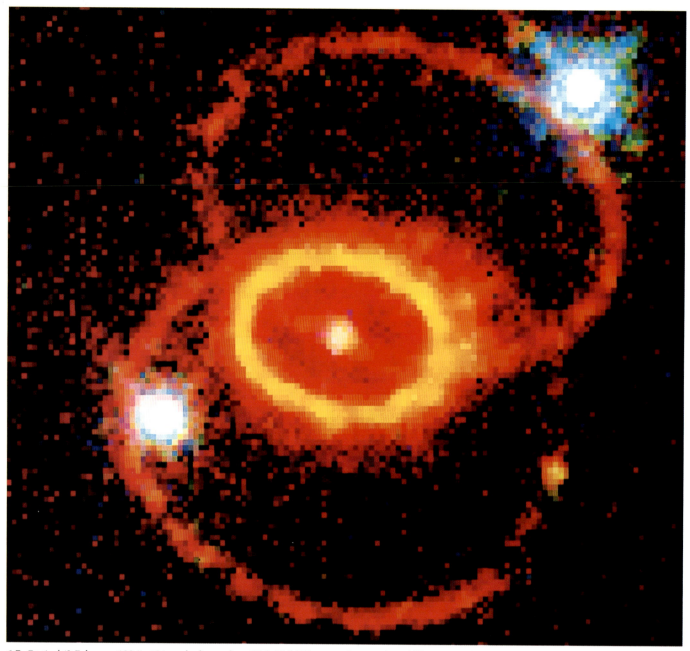

6.7 *Optical (3 February 1994), 656 nm hydrogen line, Wide Field/Planetary Camera 2, Hubble Space Telescope*

Supernova 1987A provided a cosmic beacon that illuminated the normally invisible clouds of dust and bubbles of gas in interstellar space.

Fig. 6.7 shows three rings of gas near the supernova, as revealed by the Hubble Space Telescope observing at the wavelength emitted by hydrogen atoms. In this false-colour view, dim regions are red and brighter regions yellow. The remains of the supernova, now a compact expanding ball of gas, appears as the central yellow-white blob. Two stars, unrelated to the supernova, are coded blue.

The brightest ring around the supernova is a fossil from the original star. Born 20 million years ago, this stellar heavyweight lived a short but brilliant life, eventually swelling up into a huge red supergiant.

Around 20 000 years ago, its distended outer layers drifted off into space and the star shrank to become a blue giant. This star was bright enough to be studied in its own right and even had a catalogue designation: Sanduleak –69° 202. So Supernova 1987A provided the first instance where astronomers had studied a suicidal star before its blaze of glory.

A fast but tenuous wind blew outwards from the surface of Sanduleak –69° 202, most powerfully from its poles. This wind moulded the surrounding gas, shed at the end of the star's red giant phase, into a circular ring around its equator. Gradually growing in size, this is the brightest ring in the Hubble image. Viewing it obliquely, we see the circular ring of gas as an oval.

Astronomers at the European Space

Agency have used this ring to pin down the distance to Supernova 1987A – and hence the Large Magellanic Cloud. Two hundred and fifty days after the supernova exploded, the International Ultraviolet Explorer satellite found strong emission lines from the ring, indicating that the burst of ultraviolet radiation from the explosion had just reached it. The radius of the ring must therefore be 250 light days (0.68 light years). To appear the size it does in the Hubble image (1.66 arcseconds in diameter), the ring – and hence the supernova – must lie at a distance of 169 000 light years.

The two faint oval rings in Fig. 6.7 are more of a puzzle. In three dimensions, they are probably circular, lying above and below the plane of the bright ring. They

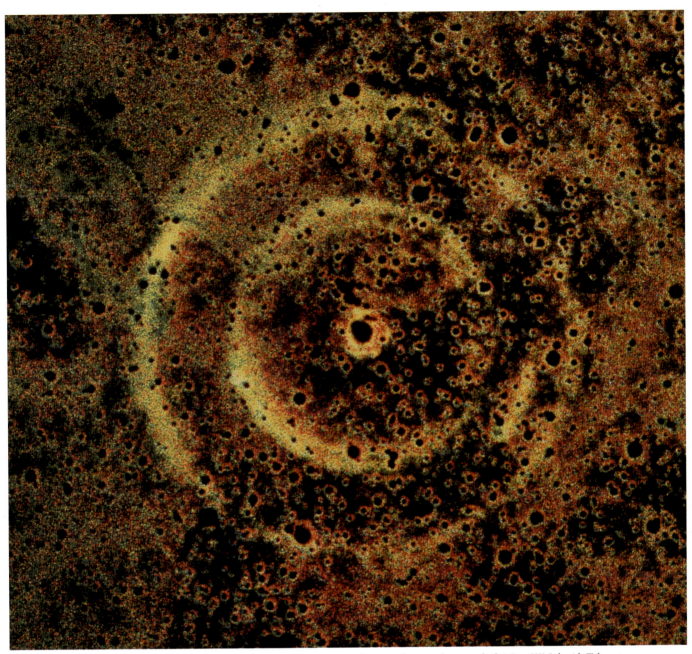

6.8 *Optical (February 1989, with February 1984 negative image superimposed), true colour, unsharp masked, 1.2 m UK Schmidt Telescope*

may be hinting at a more complex symmetry to the way that red giants shed their outer layers. Or the rings may be 'painted' on the surrounding tenuous gas by beams of energetic particles from a companion to the supernova.

The spectacular celestial bull's-eye in **Fig. 6.8** is drawn on a much larger scale: the bigger ring here is one hundred times the diameter of the bright oval in Fig. 6.7. These two rings, centred on the supernova, appeared a year after the explosion and are growing gradually larger. To see the rings more clearly, astrophotographer David Malin has 'subtracted' the images of stars and permanent nebulosity in this crowded region by superimposing a negative image taken before the supernova explosion.

These circles of light are not real rings of interstellar material. They are 'light echoes': regions of interstellar dust lit up by the supernova's brilliant light. The dust in fact lies in two curtains across our line of sight, 400 and 1000 light years in front of the supernova, and the ring shape is a geometrical trick.

If the supernova were shining with a constant light, the two curtains of dust would be lit up uniformly, like the windscreen of a car being driven towards the Sun. But the supernova explosion was, on the cosmic scale, a sudden flash of light. The light coming directly towards the Earth reached us early in 1987 and then faded. Light travelling in other directions, and redirected towards the Earth by the curtains of cosmic dust, had to take a longer route, and so reached our telescopes later.

Each part of a light echo is thus a re-enactment of the explosion, delayed by the light's roundabout path. And at any particular time, we are observing echoes shifted by the same amount from our line of sight to the supernova. As a result, the delayed light from Supernova 1987A forms a circle around the supernova's position. As time goes by, we observe light that is more and more delayed by travelling an ever longer distance before it is redirected, and so the light echoes constantly grow larger.

The light echoes are distinctly orange in this true-colour view. The supernova was yellow at its maximum brightness, and the interstellar dust has deepened this colour by absorbing more blue light – just as the setting Sun is reddened by dust in Earth's atmosphere.

6.9 *Optical, true colour, 5 m Hale Telescope*

Crab Nebula

IN THE EARLY morning of 4 July 1054, watchful Chinese astrologers saw a bright new star rising in the east just before the Sun. Over succeeding days, the 'guest star' brightened until it outshone all the other stars in the sky. For a period of three weeks it was brilliant enough to be visible in daylight; then it faded, until by April 1056 it had disappeared from sight even on the darkest night.

The 'guest star' has since been identified as a type II supernova: the explosion of a heavyweight star at the end of its life. The wreckage of the star is still expanding outwards, as the Crab Nebula (**Fig. 6.9**). It is some 6500 light years away from us, and over the nine centuries since the explosion has expanded into an egg shape about 15 light years long and 10 light years across.

English astronomer John Bevis first noticed this dim, nebulous patch in the constellation Taurus in 1731. Twenty-seven years later, it was rediscovered by the French comet-hunter Charles Messier, who put it first in his famous catalogue of nebulae. Hence the Crab is also known as M1. The great nineteenth century amateur astronomer, the third Earl of Rosse, saw for the first time 'resolvable filaments . . . springing principally from its southern extremity'. In his first drawing, the nebula's overall shape resembled the pincers of a crab – hence the nebula's popular name.

The Crab is a moderately bright nebula in visible light, but it is far from being a conspicuous object. At magnitude 8, the Crab is too faint to be seen with the unaided eye. When observed at other wavelengths, however, it ranks with the most brilliant celestial objects. As a result, it has always been one of the first objects identified when astronomers have explored new wavelengths. In 1949, Australian radio astronomers pinned down the position of the intense radio source 'Taurus A', and identified it as the Crab Nebula. The story was repeated in the early years of X-ray astronomy, by a rocket flight in 1964 which discovered and identified the Crab Nebula as an X-ray source.

Four years later, radio astronomers found that the 'power-house' of the Crab Nebula is a pulsar at its centre. This is the youngest pulsar known, one of the fastest pulsing and it was the first to be associated with a known supernova event. The Crab Pulsar is visible in this photograph as the lower of the two central stars. Its light is, in fact, flashing thirty times per second. It was the first pulsar to be detected flashing at optical and gamma-ray wavelengths.

In the late 1960s, British astronomer Geoffrey Burbidge summed up its pivotal role in the development of the new astronomy: 'there are two kinds of astronomy – the astronomy of the Crab Nebula and the astronomy of everything else'.

Tycho's supernova remnant

THE GREAT DANISH astronomer Tycho Brahe was in for a surprise as he took his early evening stroll on 11 November 1572. As he looked up, 'directly overhead, a certain strange star was seen, flashing its light with a radiant gleam'. The new star was in the constellation Cassiopeia, and night after night Tycho carefully measured its position and its brightness compared to the stars and planets visible at the same time. A few days later, the star reached its maximum brightness, shining as brilliantly as Venus with a magnitude of −4. Then it gradually faded until it became invisible some fifteen months later.

From Tycho's careful measurement of the star's fading, we can deduce that it was a type I supernova. The remnant it has left is extremely faint at optical wavelengths. Canadian astronomer Sidney van den Bergh took this optical photograph (**Fig. 6.19**) with the Hale 5-metre telescope in 1970. Even with such a large telescope and a two-hour exposure, the wisps of glowing gas hardly show. The brightest of these long, extremely thin filaments lies near the left-hand edge of the picture and runs vertically. Above and to the right, a more diffuse patch of wisps heads towards the top right, and beyond these is another isolated wisp at right angles. In this photograph through a red filter, we are seeing the light emitted by hydrogen atoms.

Most of the gas in the supernova remnant is too hot to produce light, but at a temperature of 40 000 000 K it shines brilliantly in X-rays. The picture from the Rosat X-ray observatory (**Fig. 6.20**) shows the remnant as a complete, almost circular shell of gas – appearing brightest at the outer edge because we are here looking through a greater thickness of shining gas. Each dot here is caused by a single energetic X-ray photon, focused by Rosat's mirrors onto its High Resolution Imager. This celestial ring is about 8 arcminutes in diameter, one-quarter of the Moon's apparent size. Since Tycho's remnant is about 7500 light years away from us, we can calculate that the gas shell has grown to a size of 17 light years during the past four centuries.

The radio image (**Fig. 6.21**) from the Ryle Telescope in Cambridge, England, shows the remnant in even finer detail. In this picture, the dimmest parts – mainly in the centre – are coded blue, with brighter regions green and red, shading to purple and white for the most intense radio-emitting regions at the top left of the gas shell. Unlike the X-rays, which come from the atoms of hot gas in the shell, the radio waves are synchrotron radiation caused by high-speed electrons whirling in the magnetic field trapped within the shell. In the radio view, we can see that the outer edge of the remnant appears extremely

6.19 *Optical, red light, negative print, 5 m Hale Telescope*

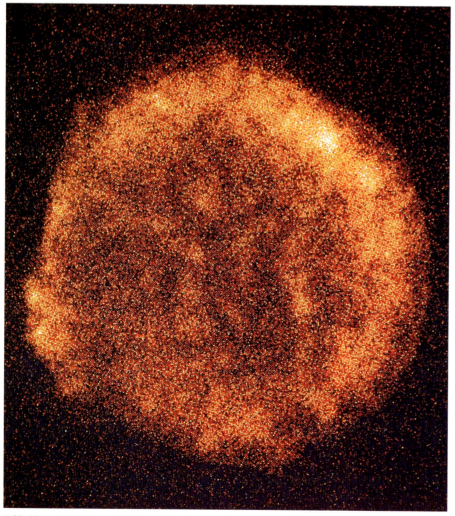

6.20 *X-ray, 0.6–10 nm, High Resolution Imager, Rosat*

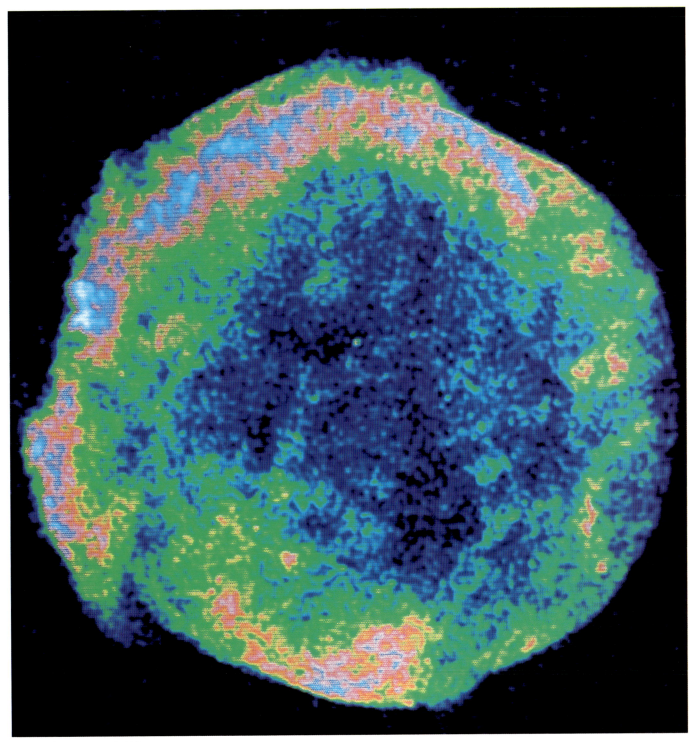

6.21 *Radio, 11 cm, Ryle Telescope*

sharp: this edge is the shock front where the shell is sweeping up fresh interstellar gas to add to its growing bulk.

The remnant's outline, seen at X-ray and radio wavelengths, is slightly 'dented', where the shell has run into small, denser clouds of interstellar gas which have slowed its expansion. The dents at the left and at the top right coincide with the two most prominent optical filaments (Fig. 6.19). They lie exactly along the edge of the shell that we see at radio wavelengths. Here the swept-up interstellar gas glows for a few years at optical wavelengths, as it is heated through a temperature of a few thousand degrees on its way to the multi-million degree temperature it will reach within the shock front.

Tycho's supernova remnant is clearly very different from the Crab Nebula (Figs. 6.9–6.18). Not only is it very faint when photographed at optical wavelengths, but the X-ray and radio emission come from a hollow shell of gases, while the Crab's radiation of all wavelengths is strongest at the centre. Whereas the Crab Pulsar powers the Crab's bright synchrotron nebula, there is no powerful radiation and energy source at the centre of Tycho's remnant. A young neutron star would be hot enough to show up as a bright point in the X-ray picture, even if its beamed radiation were emitted in a direction which missed us. This remnant is powered entirely by the momentum of the supernova explosion.

The subtle shades in colour in the composite optical photograph (**Fig. 6.24**) show that different parts of the Cassiopeia A remnant are made of different mixtures of gases, and prove that the exploding star has thrown out newly created elements into space.

Sidney van den Bergh and Karl Kamper prepared the false-colour composite by combining three photographs, taken in the mid-1970s through different coloured filters, to isolate the light from only one or two types of atom on each photograph. The original negatives have been printed with different coloured illumination onto the same positive print, in colours corresponding to wavelengths shorter than those of the original filters.

The extensive 'reddish' nebulosities visible here are emitting light from sulphur or oxygen atoms at long red wavelengths, around 670 nanometres. The 'green' patches of nebulosity are producing light from hydrogen and nitrogen, in reality at shorter, but still red, wavelengths of 656 nanometres. The faint 'bluish' filaments (on the left) shine in the green light of oxygen at 501 nanometres.

The different coloured filaments are moving at different speeds, and this gives valuable clues to the history of the supernova explosion. The 'greenish' ones are the slowly moving 'quasi-stationary flocculi'. Some thirty of these can be seen on the original plate, and they travel so slowly and alter so little that these flocculi can be recognised at almost the same positions in the photograph taken some twenty years earlier (Fig. 6.22).

The more intense patches of 'reddish' nebulosity, and the few 'blue' filaments are fast-moving knots. About a hundred distinct knots are visible on the original photographs, but they are moving and changing so rapidly that it is difficult to recognise any individual filament after a few years have passed. For example, the short arc of 'reddish' nebulosity on the lower right of the photograph has appeared only recently. The earlier photograph (Fig. 6.22) does not show it at all, although the 'green' quasi-stationary flocculus next to it can be easily picked out.

These fast-moving knots show no sign of light from hydrogen (which would appear green in this representation), the most common element in the Universe and the gas which makes up most of the outer layers of ordinary stars. Instead, they shine in the light of oxygen, sulphur and argon.

6.24 *Optical, 501 nm oxygen line (blue), 656 nm hydrogen line (green), 670 nm sulphur line (red), 5 m Hale Telescope*

The unusual composition of these knots suggests that they are 'lumps' of gas from deep inside the original star, thrown out into space in the supernova explosion. During its lifetime, a star converts hydrogen at its core into helium, then the helium into heavier elements like carbon, oxygen, sulphur and argon. The knots in Cassiopeia A consist of exactly the same gases we would expect to find near the centre of a star many times heavier than the Sun – and theoretical studies tell us that such a heavy star should end its life as a supernova.

The gases emitting visible light are, however, only a small fraction of the total thrown out by the star, and swept up from interstellar space by the expanding remnant. The shock of this sweeping-up process heats the gases to temperatures of millions of degrees, and such gas emits not light, but X-rays. The X-ray view from the Einstein Observatory (**Fig. 6.25**) thus shows where most of the matter lies. Although it looks very similar to the radio photograph (Fig. 6.23) the X-ray structure is a result not of magnetic fields creating synchrotron radiation, but of radiation from the dense, superhot gases.

The observatory's High Resolution Imager was used to resolve the fine details of this small X-ray source. Each dot corresponds to a single X-ray photon impinging on the detector.

Fig. 6.25 shows two concentric shells making up this supernova remnant. The outer shell is formed by the swept-up interstellar gases. These gases are at a similar temperature to those in the swept-up shell of Tycho's supernova remnant (Fig. 6.20), around 50 million degrees. The inner, more intense shell on the X-ray photograph is rather cooler – 'cool' here meaning some ten million degrees – and the gas is also much denser. This inner shell contains most of the matter thrown off in the star explosion – about fifteen solar masses of gas. This gas from deep within the star has been travelling outwards at thousands of kilometres per second, but it has lagged behind the more tenuous gases from the star's surface, which shot ahead to form the outer shell. As the gas in the outer shell sweeps up interstellar gas, it is slowing down, and the inner shell is catching up with the outer shell. The collision, or 'reverse-shock', is heating up the inner gas shell. Because it is so massive, the inner shell emits X-rays copiously and dominates this view.

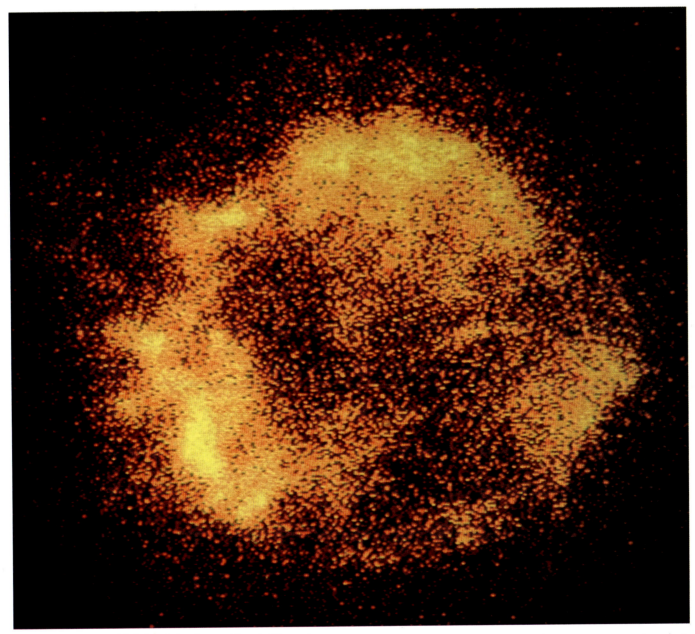

6.25 *X-ray, 0.4–8 nm, High Resolution Imager, Einstein Observatory*

This 'complete view' of Cassiopeia A (**Fig. 6.26**) combines three pictures which span the entire wavelength range from X-rays (green), through optical (red) to the long-wavelength radio waves (blue). The 'green' X-ray image is the most fundamental, revealing most of the matter from the deceased star – some fifteen solar masses. The 'red' optical view shows many stars; the faint optical knots and flocculi of Cassiopeia A contain less than one per cent of its gases. The 'blue' radio image reveals the tangled magnetic fields.

The composite exposes the relations – and contrasts – between these aspects of the explosion. On the broad scale, the radio (blue) image is brightest to the right, and the X-ray (green) to the left. American astronomer John Dickel, who made this composite, suggests two possible causes. Perhaps the magnetic field in interstellar space is slightly stronger to the right, and when swept up its interaction with the electrons has produced stronger radio synchrotron emission. Or the gas in interstellar space between Cassiopeia A and the Sun is uneven, and denser gas to the right absorbs X-rays more efficiently.

Most of the optically bright fast-moving knots lie at the North (top) of Cassiopeia A. The radio and X-ray images are intense here too, but their brightest parts do not coincide exactly with the optical knots. Nonetheless, they are probably related. The knots are dense blobs of gas, travelling outwards from the original explosion. As they enter the inner shell of gas, they churn up its magnetic field and so create stronger synchrotron radio waves from the region around.

The knots are comparatively cool, however, and the gas shell at a temperature of 10 000 000 K rapidly heats them up. The outer parts of the knots thus 'evaporate' to join the rest of the gas in the shell. This recently evaporated gas is still fairly dense, and hence emits X-rays powerfully. The result is that many of the optical knots are surrounded by a 'halo' of intense X-ray emission. Small, bright X-ray clouds without an associated optical knot may be regions where one of them has recently evaporated completely.

The quasi-stationary flocculi, seen at optical wavelengths, lie further out, in the interstellar gases just beyond the shock front (marked by the outermost edge of the remnant as seen in radio and X-rays). Many astronomers believe that the flocculi are small dense gas clouds which drifted away from the original star long before it exploded. Now the shock wave from the explosion is catching up with them. As the shock engulfs a flocculus, it glows at optical wavelengths. Parts of an engulfed flocculus are heated to multi-million degree temperatures. They shine as brighter X-ray spots in the generally faint outer shell, appearing adjacent to optical flocculi.

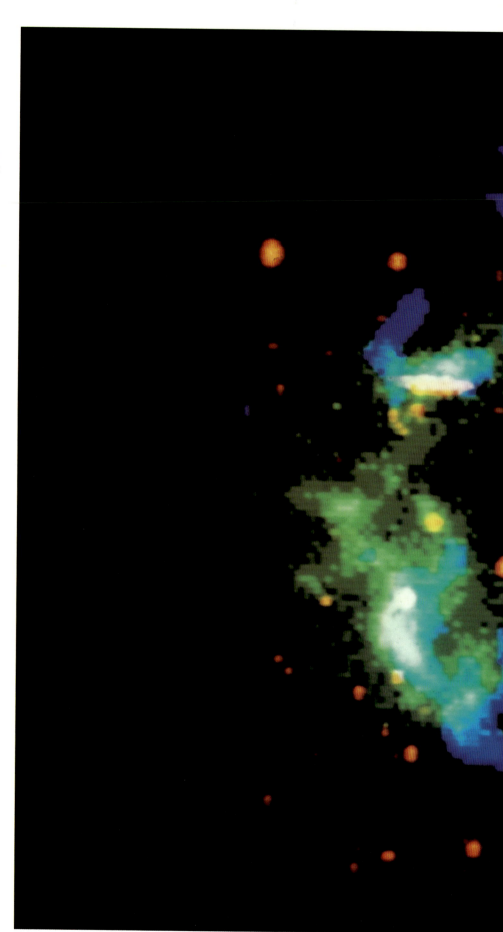

6.26 *Combined wavelengths: optical (red), red light, 5 m Hale Telescope; radio (blue), 11 cm National*

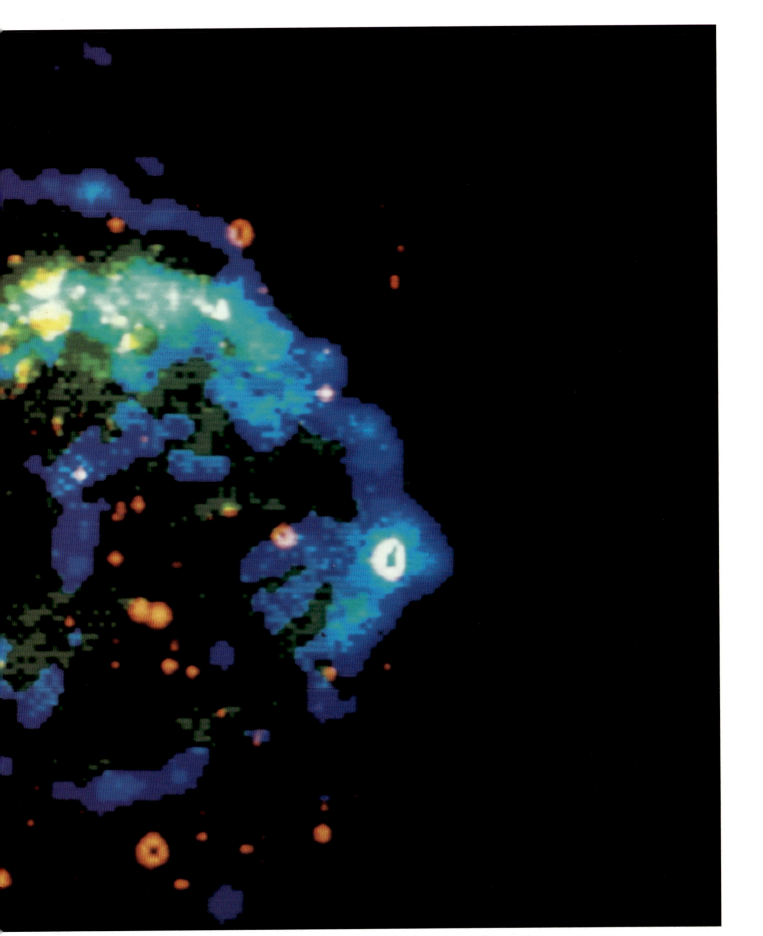

Radio Astronomy Observatory interferometer; X-ray (green), 0.4–8 nm, High Resolution Imager, Einstein Observatory

Vela supernova remnant

AROUND 10 000 BC, a supernova erupted near to the Sun, in the southern constellation of Vela. For a whole month, it was more brilliant than the Moon: the supernova was visible even in daylight.

By now, the wreckage from this exploding star has spread to a size of 140 light years. Lying only 1500 light years from us, the supernova remnant fills several degrees of sky. Optical telescopes can detect little sign of it, but at other wavelengths the Vela supernova remnant is among the most brilliant celestial objects. The images on this page reveal its appearance in four very different wavelength bands.

The optical photograph (**Fig. 6.27**) was taken through a green filter to emphasise the light from glowing gas, and dim the background of stars. A negative print helps to reveal the faintest wisps. These delicate traceries are thin sheets of glowing gas, draped in folds around the expanding ball of debris. Where a sheet runs across in front of us, its light is spread out and it fades to invisibility. But where we view a sheet edge-on – at the side of the remnant or at a fold – its light appears concentrated into a narrow bright filament.

The filaments around the outside of Fig. 6.27 mark the edge of the supernova remnant. The glowing gas at the top left has been dimmed by a dark cloud in the foreground, which has also extinguished distant stars to make this part of the negative print unusually pale. The bright filament down the centre – shining through a gap in the obscuring clouds – is a crease in the front part of the expanding ball of debris. These thin sheets reveal where the shock front from the explosion is driving outwards through the interstellar gases, and lighting them up as it passes.

The shock heats the swept-up gases to over a million degrees, so the material within the shock shines brilliantly at very short wavelengths. In the extreme ultraviolet (between ultraviolet and X-ray) the Vela supernova remnant is among the brightest sources in the sky.

Fig. 6.28 is the view from the Extreme Ultraviolet Explorer satellite. It is colour coded for intensity, with the brightest regions white, and successively dimmer parts pink and blue. At first glance, two bright patches stand out. The region between them has been dimmed by the same dark cloud that obscures this region in the optical view. But faint emission is visible here, and also on the lower right, forming a complete, though uneven, ring.

In three dimensions, this apparent ring of hot gas forms a hollow shell. It consists almost entirely of swept-up interstellar gas: over the millennia, the expanding remnant has swept up far more gas than was thrown off the original supernova.

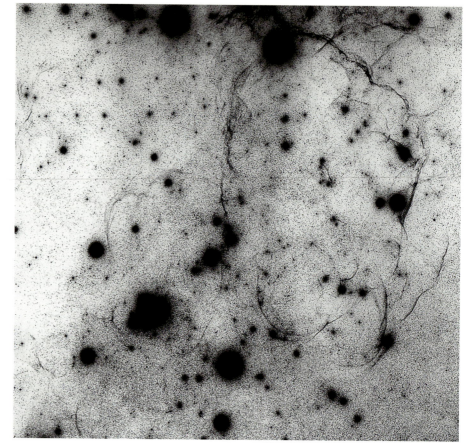

6.27 *Optical, 501 nm oxygen line, negative print, 1.2 m UK Schmidt Telescope*

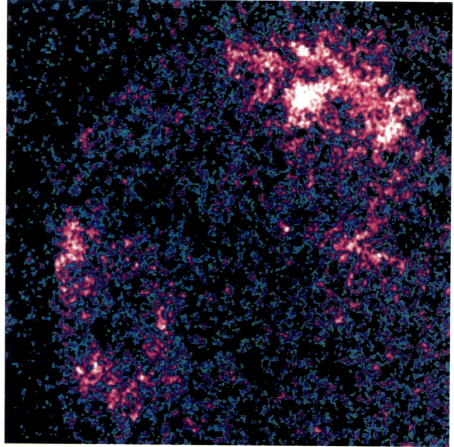

6.28 *Ultraviolet, 5–18 nm, Extreme Ultraviolet Explorer*

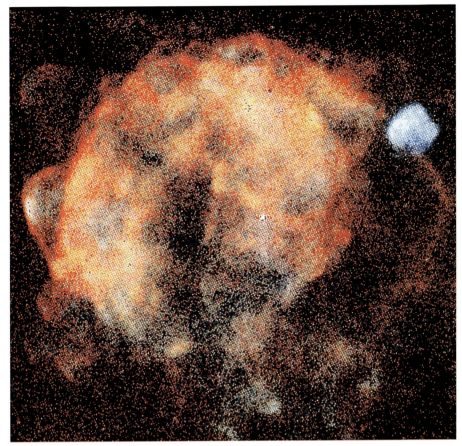

6.29 *X-ray, 0.6–10 nm, Position Sensitive Proportional Counter, Rosat*

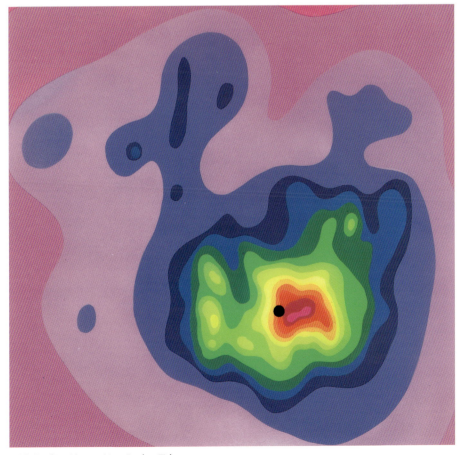

6.30 *Radio, 11 cm, 64 m Parkes Telescope*

At X-ray wavelengths, the Vela supernova remnant is a stunning object. **Fig. 6.29** covers a slightly larger area of sky (to the right and below), and shows what we would perceive if our eyes were sensitive to X-rays: the brightness corresponds to X-ray intensity and the colours to wavelength. As a result, red indicates regions of cooler gas and blue the hotter parts. Observing at these high energies, even 'cooler' means a million degrees!

The bright blue patch (upper right) is another supernova remnant lying in the background. Four times further away than Vela, the small Puppis A remnant is only 4000 years old. Its gases are still expanding so fast that they heat the swept-up interstellar gas to ten million degrees.

Vela contains gas at a variety of temperatures. The lower part of the remnant has encountered little resistance, and the gases (blue) are at eight million degrees. But the rest of the expanding shell has swept up so much interstellar material that the shock wave has slowed down and lost much of its power. The gas here is at only 900 000 K (red).

This X-ray image was the first to show the faint extension to the right, beyond the optical filaments, and also some small blobs scattered around the edge of the remnant. These are probably 'bullets' of dense gas ejected at high speed from the supernova's core. The site of the explosion is marked by the tiny bright source in the middle of Fig. 6.29: it is a compact neutron star, the Vela Pulsar.

This small powerful object is the clue to the radio appearance of the Vela supernova remnant, **Fig. 6.30**. It covers the central two-thirds of the region seen in the previous three images, and is colour coded so that the faintest regions are pink, with successively brighter parts in shades of blue, green, orange and red. The Vela Pulsar is marked by the black dot.

At first glance, the remnant looks entirely different – its radio appearance has led to the nickname the Donald Duck Nebula! In fact, the dim outer regions of the radio source (catalogued as Vela Y and Vela Z) correspond to parts of the shell seen in the short-wavelength images.

The bright central region of the radio image (green, yellow and red) – Vela X – lies in the centre of the hollow supernova remnant. Here there is very little gas. The radio emission comes instead from the synchrotron emission of high-speed electrons in a magnetic field. Both the electrons and the magnetism are generated by the Vela Pulsar.

In fact, Vela is a composite supernova remnant. It has an outer shell like Cassiopeia A, where interstellar gas is being swept up, and also an interior synchrotron nebula (Vela X) that is very like the Crab Nebula but five times larger.

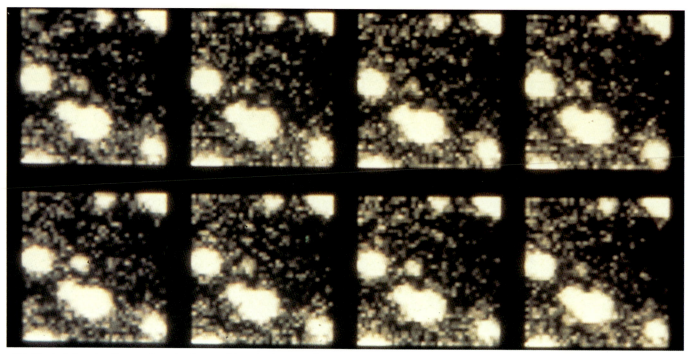

6.31 *Optical, Imaging Photon Counting System, 3.9 m Anglo-Australian Telescope*

Right at the heart of the Vela supernova remnant lies the tiny, rapidly spinning Vela Pulsar. Australian researchers discovered radio pulses from this part of the sky in 1968, and optical astronomers searched for corresponding flashes of light (**Fig. 6.31**).

This sequence of eight frames shows the pulsar going through one complete cycle. The series runs horizontally from top left to top right, and then from bottom left to right. The bright spots are ordinary stars, while the pulsar is in the middle of each frame. It flashes twice in each cycle (in the first and third frames of the lower row). At the time these pulses were detected, the Vela Pulsar was the faintest optical object ever observed, and only the second pulsar found to emit light (after the Crab Pulsar).

Like other pulsars, Vela is a cosmic lighthouse. It seems to flash as a rapidly spinning neutron star sweeps narrow beams of radiation across the Earth. Oddly enough, the two optical flashes are not half a cycle apart, and neither coincides with the radio pulse, which occurs in the third frame of the top row. So Vela is not producing just one beam of radiation, or two in opposite directions (like the Crab). The Vela Pulsar has at least three beams – two of light and one of radio waves – pointing in different directions.

The entire sequence in Fig. 6.31 lasts less than one-tenth of a second. At the time of these observations, it was the fastest known repetition rate apart from the very young Crab Pulsar. The pulses are slowing down at four millionths of a second in a year. Working backwards, astronomers calculate that the pulsar was born 12 000 years ago. This method provides our best estimate of the date of the Vela supernova.

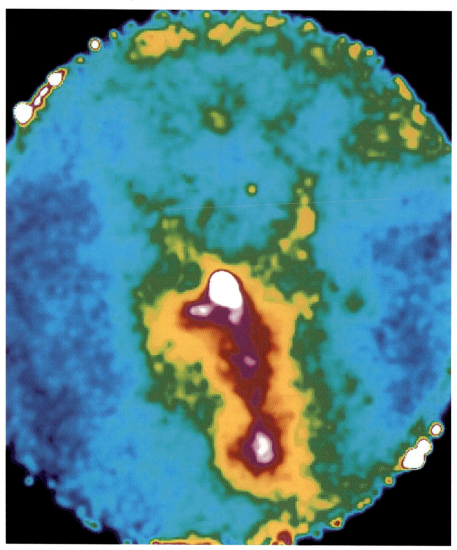

6.32 *X-ray, 0.6–1.4 nm, Position Sensitive Proportional Counter, Rosat*

6.33 *Optical, true colour, 1.2 m UK Schmidt Telescope*

As the pulsar spins down, it must be losing energy at a great rate. In 1995, X-ray astronomers found just where this energy is going. **Fig. 6.32** is an X-ray image of one degree of sky (twice the apparent size of the Moon), centred on the Vela Pulsar. Colour coding shows X-ray brightness, from blue for the dim background, through green, yellow, red and pink to white for the most intense regions.

The pulsar is the brilliant white spot at the centre. What surprised astronomers is the broad jet (red) of hot gas flowing from the pulsar, towards the lower right, at a speed of 1000 kilometres per second. This gas is being picked up by superfast electrons beamed out from the pulsar. At the end of the jet, the electrons diffuse outwards, broadcasting their energy as the radio waves coming from Vela X (Fig. 6.30). The stream of electrons and gas pushes the pulsar in the opposite direction: the Vela Pulsar is, literally, jet-propelled! If its inbuilt jet keeps firing for long enough, it will propel the pulsar outside the supernova remnant and off into the outskirts of our Galaxy.

Meanwhile, the gaseous supernova remnant will slow down as more interstellar gas is snowploughed up, and its temperature will fall. Instead of copious X-ray emission, cooler filaments emitting the green light of oxygen and red hydrogen light (**Fig. 6.33**) will become more common. They radiate away energy much more efficiently, and the gas will quickly condense into small dense clouds. From the ashes of this one exploding star will arise, phoenix-like, a whole new cluster of stars.

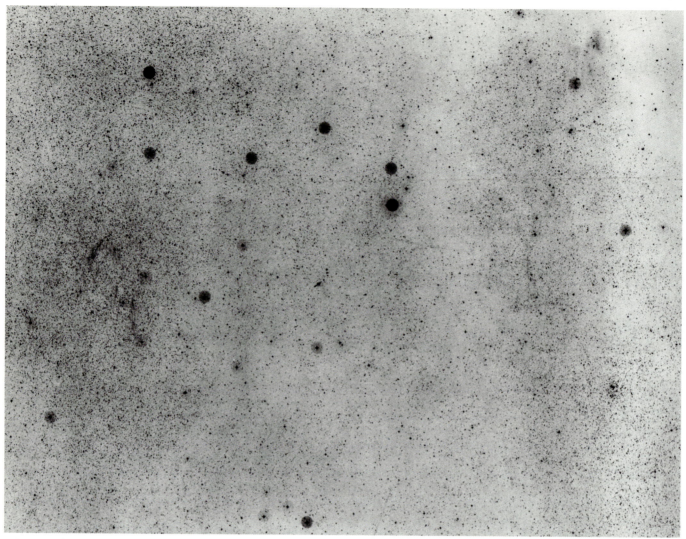

6.34 *Optical, negative print, 1.2 m Palomar Schmidt Telescope*

SS 433

THE ARROWED STAR in the photograph (**Fig. 6.34**) looks misleadingly unimpressive – similar to the hundreds of others here. In fact, it is the notorious SS 433, one of the most bizarre objects in our Galaxy.

SS 433 achieved fame in 1978, but it had been noticed in the 1960s, when American astronomers Bruce Stephenson and Nicholas Sanduleak were looking for stars with peculiar spectra. This faint (magnitude 14) star in Aquila had bright spectral lines of hydrogen and duly went down as number 433 in the Stephenson-Sanduleak ('SS') catalogue. In 1978, Canadian astronomer Ernie Seaquist searched for radio waves from SS stars, and found that SS 433 is a radio emitter. But Seaquist did not realise that SS 433 is also surrounded by a huge ring of radio emission, known as W50 from its cataloguing by Dutch radio astronomer Gart Westerhout in the l950s.

Radio astronomers in Australia and Cambridge were investigating W50 at the time, and they independently discovered the intense radio source at the centre. It turned out to be an X-ray source too. Intrigued by these results, optical astronomers David Clark and Paul Murdin pointed the Anglo-Australian Telescope at this position – and rediscovered the star with the strong hydrogen lines. Later studies of SS 433's hydrogen lines showed that they are moving regularly backwards and forwards in wavelength by a small amount, due to the Doppler effect, indicating that this is a double star system, where the stars orbit one another with a period of 13 days.

One of the pair is a fairly normal star, whose outer gases are falling onto a compact companion star – a neutron star – in a spiralling accretion disc at a rate of a hundred Earth masses per year. This whirling disc's hot gases generate the X-rays, and its magnetic field generates the radio waves which we pick up from SS 433. Its light comes from the cooler, outer regions of the accretion disc and from the normal star. So far, the system resembles other X-ray binaries, like Scorpius X-1. Clark and Murdin had, however, noticed other bright lines in the spectrum of SS 433, at wavelengths which are not emitted by any of the common elements. American astronomer Bruce Margon followed up these observations, and discovered that these unusual lines were changing dramatically in wavelength, first one way then the other. They were marching up and down the spectrum in a regular rhythm which repeats every five-and-a-half months.

The lines are moving back and forth around the wavelengths of the ordinary hydrogen lines, so they are simply due to clouds of hydrogen, and the wavelength change must be a tremendous Doppler shift. What we are seeing is gas in two 'jets' 'boiled off' either side of the hot accretion disc, and swinging round with a period of five and a half months as the disc precesses, like an unstable top.

The jets themselves keep a constant speed. SS 433's great surprise was the enormous size of this speed: just over a quarter of the speed of light – some 80 000 kilometres per second. Although electrons are travelling at such high speeds

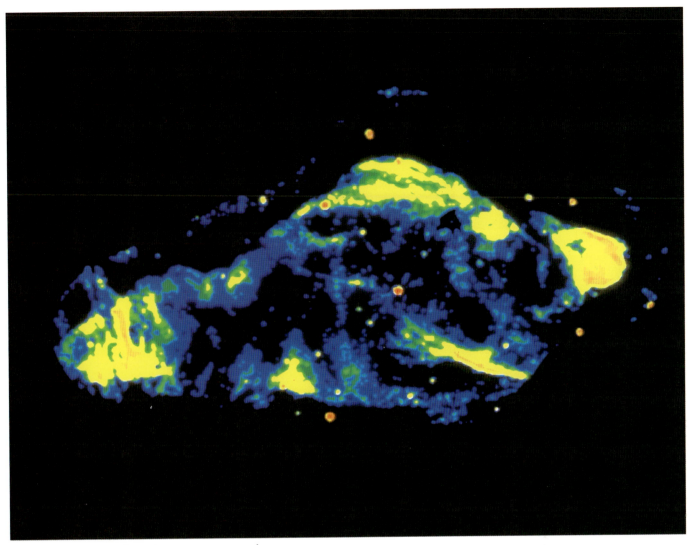

6.35 *Radio, 20 cm, Very Large Array*

in many cosmic radio sources, the discovery of the speed of these jets of ordinary gas was totally unexpected.

In the radio picture (**Fig. 6.35**), covering a slightly larger area than the optical photograph, the most intense radio regions are colour coded red and orange, with successively fainter regions yellow, green and blue.

SS 433 itself is the red spot at the centre while W50 is the distorted oval filling most of the picture. The radio waves from W50 come from a thin, hollow, egg-shaped shell, where fast electrons, whirled around in compressed magnetic fields, generate synchrotron radiation. It stretches about $2/3°$ North–South and $2°$ East–West (four Moon breadths). W50 is about 18 000 light years away, which means that it is huge for a supernova remnant: 200 light years by 600 light years in extent.

Apart from its size, W50 differs from other supernova remnants in its shape. It is neither circular in outline, like Tycho's remnant, nor oval like the Vela remnant. The easiest way to describe W50 is as an oval with 'ears'. In Fig. 6.35, the bright

yellow and orange region to the right is the smaller ear, while the larger ear to the left has almost broken away from the central shell. More detailed radio observations (Fig. 6.36), and X-ray pictures (Fig. 6.38), reveal that the jets from SS 433 are heading out East and West (left and right), and it seems that they are pushing back the interstellar gas there with more force than the expanding supernova shell can muster, and so are inflating the radio ears of W50.

The radio map shows no distinct traces of the 'inner edges' of the ears, where the oval rim of the supernova remnant would have run in their absence, but these do show in the optical photograph (Fig. 6.34) as thin wisps of gas near the left and right edges.

The simplest explanation of the formation of the SS 433/W50 complex is that two very similar stars were born together just over five million years ago, each with a mass of as much as 25 Suns. The slightly heavier, shorter-lived star exploded as a supernova after only five million years, and its core was left as the neutron star in SS 433. Now, 100 000

years later, we detect this supernova's remnant as W50, while the overspilling gases from the other star are powering the radiation and fast jets of SS 433.

Dutch astronomer E. P. J. van den Heuvel, however, suggests an alternative scenario: the two original stars were about twenty and eight times heavier than the Sun, and they were born ten million years ago. The heavy star exploded as a supernova three million years ago, and its supernova remnant has long since dissipated into space. Some 40 000 years ago, however, the lightweight star started to lose gases to the compact star left over from the explosion, which then began to squirt out its high-speed jets of gas. The jets are entirely responsible for blowing up the 'balloon' of W50, which is thus not a genuine supernova remnant at all.

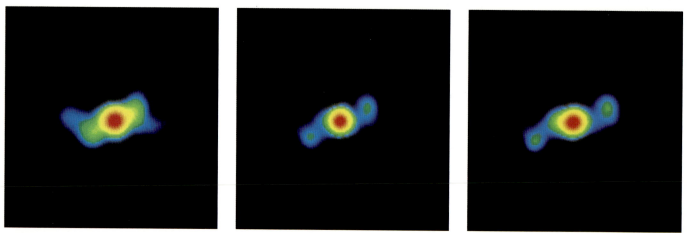

6.36 *Radio (January, February, March, April 1981), 6 cm, Very Large Array*

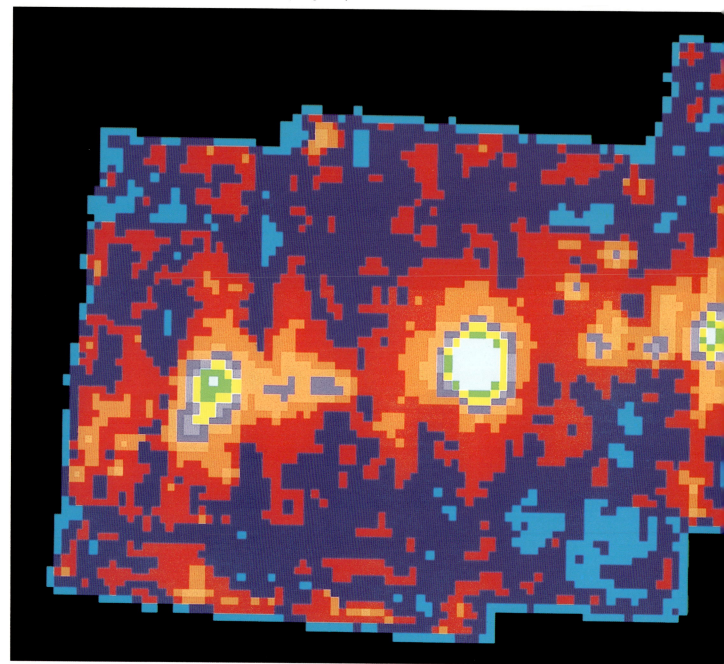

6.38 *X-ray, 0.3–2.5 nm, Imaging Proportional Counter, Einstein Observatory*

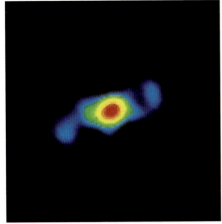

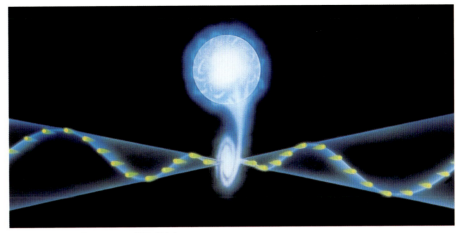

6.37 *Diagram, formation of SS 433 jets*

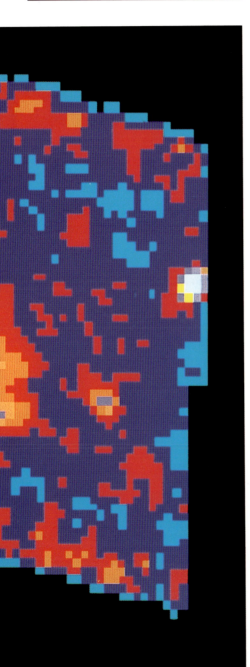

Detailed radio observations (**Fig. 6.36**) reveal that the first interpretations of the optical spectrum were astonishingly successful in interpreting the three-dimensional structure of SS 433 and its jets. Israeli astronomer Mordehai Milgrom, and George Abell and Bruce Margon in the United States, had worked out from the spectrum that there must be two oppositely directed jets, swinging around like the spray from an eccentric lawn sprinkler, tipped up as shown in the diagram (**Fig. 6.37**). The axis of the 'sprinkler' runs more or less across the sky, at an angle of about 80° to our line of sight. Each jet swings around in such a way that it is always at an angle of 20° to this axis, so the jet is pointing along one of the two imaginary cones.

The jets consist of successive 'blobs' of gas shot out by SS 433's accretion disc, each blob following its own path in a straight line along the direction in which it was originally ejected. If we were to draw a line to connect the successive blobs in each jet, it should make an unwinding spiral along one of the imaginary cones.

With the Very Large Array radio telescope in New Mexico, astronomers then detected the blobs making up SS 433's jets – and by observing them month after month they found that the jets behave exactly as Milgrom, Abell and Margon had predicted. The four observations here (Fig. 6.36) were made in 1981, at intervals of about a month from mid-January to mid-April. The pictures are colour coded so that the brightest regions are red, with successively fainter regions yellow and green, and the dimmest parts blue.

Three-quarters of the radio emission comes from SS 433 itself, the central red blob. But the two jets clearly emit radio waves too. They stretch out about one-sixth of a light year on either side of SS 433 (these maps are at a scale 250 times larger than the previous radio image of SS 433 and W50, Fig. 6.35). The first picture shows a pair of outer blobs (blue) at about positions of 'three o'clock–nine o'clock';

by the second view these have faded from sight. In the first picture, there is also an inner, more intense pair of blobs (green), emitted at a different angle ('two o'clock–eight o'clock') because the accretion disc has swung around. In the later pictures we can see this pair of blobs moving steadily outwards along the same directions, and gradually fading (to blue) as they go. By the last frame, they have almost disappeared. Now we can see the next pair of blobs, closer in at 'three o'clock–nine o'clock'.

The composite of three views from the Einstein Observatory (**Fig. 6.38**) is about 1° across, half the extent of the radio image of W50 (Fig. 6.35). The colours run from white for the most intense X-ray emission at the centre (from SS 433 itself) down through green, yellow, purple, orange and red, to purple (again) and blue for the faintest regions.

The X-ray jets here stretch out a hundred light years in each direction, about six hundred times further than the radio jets above. They show how the 'cones' fan out from SS 433, and reach the inner edge of the radio ears of W50, where the optical nebulosity lies (Fig. 6.34). The X-rays may come from hot gas in the blobs, or from synchrotron radiation like the radio jets; in either case they show that the jets have been travelling outwards for a thousand years or more.

As well as the main jets running roughly East–West, there are other X-ray blobs around SS 433 in this view. Some of them form pairs that straddle SS 433 – at 'seven o'clock–one o'clock', for example, and particularly the line of compact blobs at 'eight o'clock–two o'clock'. Has SS 433 shot out its two jets at completely different orientations in the remote past, to create these other pairs of blobs? This might explain some of the strange curved streaks in the radio image of W50 (Fig. 6.35). Many questions still haunt astronomers caught up in the mystery of SS 433.

G109.1-1.0

IS THIS THE missing link between the bizarre SS 433 and ordinary supernova remnants? G109.1-1.0 is a large semicircular ring of gases in the constellation Cassiopeia (Fig. 6.1), shining brightly in radio waves and X-rays, but only faintly as seen in optical telescopes (**Fig. 6.39**).

It is a rather aged, misshapen supernova remnant. But a 'star' in the centre is apparently ejecting jets of gas highly reminiscent of SS 433's jets. Canadian astronomers Phil Gregory and Greg Fahlman discovered this extraordinary supernova remnant by its X-ray emission, when they observed this part of the sky with the Einstein Observatory in 1979. In the X-ray image (**Fig. 6.40**) we see radiation from hot gases in the remnant. The left-hand rim shines brightly (coded yellow) while the right-hand side is much fainter (blue). But the real surprise is the very bright (red) compact X-ray source to the right of the centre.

The X-rays from this compact source pulse regularly with a period of seven seconds. Relatively slow X-ray pulsars like this are invariably part of a double star system where one partner is a rotating neutron star, feeding on gases from the other, normal star. The neutron star is presumably the core of the supernova which created the surrounding remnant.

In the radio image (**Fig. 6.41**), the brightest regions are coded red and fainter parts yellow and green, while the background sky is blue. The remnant is 33 by 25 arcminutes in extent (about the same apparent size as the Moon). Its radio emission is synchrotron radiation from fast electrons moving through a magnetic field. G109.1-1.0 lies 12 000 light years from us, so that its shell is, in fact, 100 light years across, suggesting that G109.1-1.0 is the remnant of a star which exploded some 20 000 years ago.

The original supernova exploded at the edge of a dense cloud of gas (a molecular cloud), which extends from the middle of these images towards the right. The small intense radio source to the right of the semicircular remnant is a nebula within the cloud, where stars have just been born. This nebula also shows up as the brightest wisp of gas on the optical photograph (Fig. 6.39) which covers the same region of sky. The dense cloud has prevented the ejected supernova gases from travelling far to the right; but they have billowed out to the left to create a hemispherical supernova remnant – which we can see in projection as a semicircle.

Gregory and Fahlman suggest that the pulsar can account for the appearance of the radio emission, which resembles two bright ovals (ellipses). According to their model, the central pulsar is emitting two jets of matter in the opposite directions,

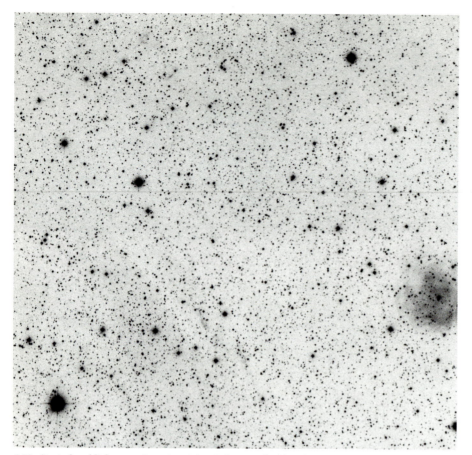

6.39 *Optical, red light, negative print, 1.2 m Palomar Schmidt Telescope*

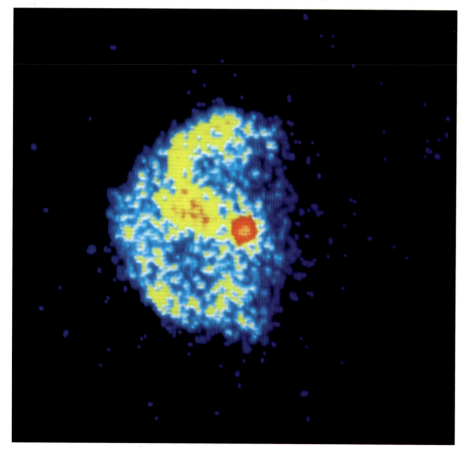

6.40 *X-ray, 0.3–2.5 nm, Imaging Proportional Counter, Einstein Observatory*

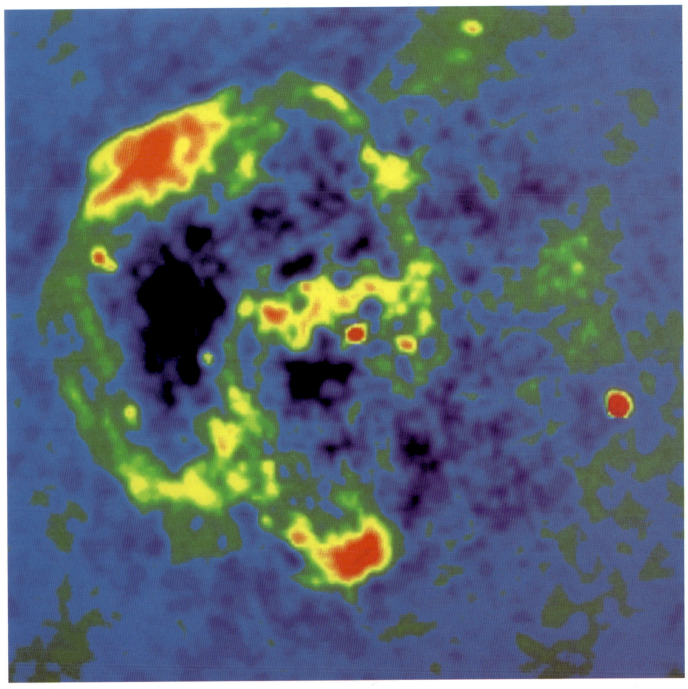

6.41 *Radio, 20 cm, Very Large Array*

like a weaker SS 433. The jets travel freely within the relatively empty cavity of the remnant's interior, but where they hit the shell of the remnant the jets generate radio waves. They swing round, like those of SS 433 but through a larger angle, and so each jet traces out a loop on the shell, over and over again. The jet travelling to the upper left draws out a large ellipse on the hemispherical part of the remnant, which happens to coincide mainly with the left-hand edge of the remnant as seen from Earth. The second jet, travelling to the lower right, draws out a smaller ellipse. The small, bright radio source here, incidentally, is not the pulsar,

which does not appear in this image.

To support this interpretation Gregory and Fahlman point out that the X-ray picture (Fig. 6.40) shows a broad jet heading from the pulsar towards the top left – the predicted direction for one of their jets – just as the jets of SS 433 can be seen as broad X-ray emitting features a long way out from the central source.

G109.1-1.0 is, however, weak compared to SS 433. Although the central sources emit about the same power in X-rays, SS 433 shines a thousand times more brightly than the pulsar in G109.1-1.0 in visible light, and it is over 3000 times brighter at radio wavelengths. In addition,

there is no *direct* evidence in G109.1-1.0 for the kind of extremely high speeds found in SS 433's jets. Nonetheless, further study of this source will undoubtedly help us understand how these strange swinging-jet star systems operate, and how they arise in the process of star death.

7 radio astronomy

IN AN AMERICA hit by the Depression of the early 1930s, scientists achieved a major breakthrough in the study of the Universe: they acquired the first view of the skies at wavelengths of invisible radiation. The new instrument was strange by any standard, and it was certainly a far cry from anyone's idea of a 'telescope'. Sited in the midst of the potato fields of New Jersey, the first radio telescope consisted of eight large metal hoops, supported on a wooden frame which rotated slowly on a set of Model T Ford wheels. It was as important a step forward as Galileo's first optical telescope of three centuries earlier. Its descendants have not only revealed weird and unexpected radio sources in the sky – pulsars and quasars, to name but two – but have shown astronomers that they must turn away from ordinary optical astronomy if they are to investigate some of the strangest, and some of the most important, objects in the sky. The spectacular growth of the 'new astronomies' – X-ray, infrared, ultraviolet and gamma ray – has been inspired by the unpredicted successes of radio astronomy.

The first instrument to detect radio waves from space was not in fact a radio telescope. The Bell Telephone Laboratories were studying sources of radio 'static' or 'hiss' which interfered with ship-to-shore communications, and a

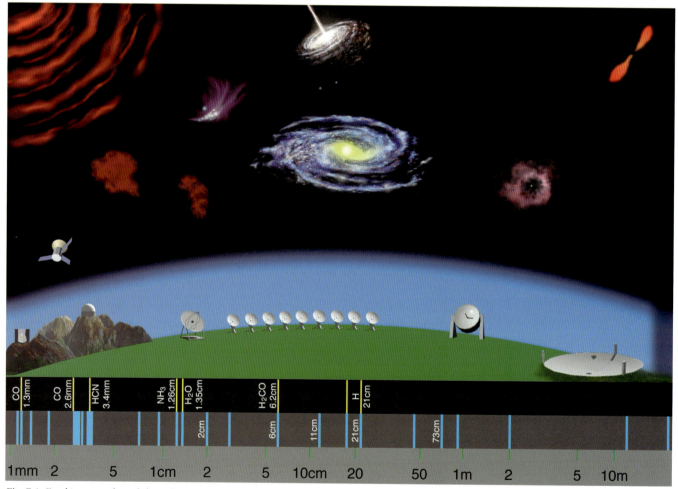

Fig. 7.1 *Earth's atmosphere defines the 'radio window' on the Universe. Gases in the lower atmosphere absorb wavelengths shorter than 1 millimetre, while the ionosphere reflects away radiation longer than 30 metres. The whole of the radio spectrum is also 'polluted' by radio, television and other artificial broadcasts, so certain wavelengths (blue) are kept free by international agreement for radio astronomy. These are spread throughout the spectrum, with some tied in to the wavelengths naturally broadcast by certain atoms and molecules, such as carbon monoxide (CO), hydrogen cyanide (HCN), ammonia (NH_3), water (H_2O), formaldehyde (H_2CO) and hydrogen (H).*
Radio telescopes for the shorter radio waves must be sited on mountain tops, such as the James Clerk Maxwell Telescope on Mauna Kea, Hawaii, and the pioneering millimetre telescope on Kitt Peak, Arizona. They tune into molecules in dark interstellar clouds. Above the atmosphere, the Cosmic Background Explorer satellite observes short radio waves from the Big Bang. Telescopes on the ground include single dishes at Effelsberg (Germany), Jodrell Bank (UK) and Arecibo (Puerto Rico), and the Very Large Array in New Mexico. They can observe the 21-centimetre emission from hydrogen in the Milky Way (upper centre), or a wide range of wavelengths from nebulae, quasars, supernova remnants and double-lobed radio galaxies.

Fig. 7.2 *The 76-metre Lovell Telescope at Jodrell Bank dominates the surrounding Cheshire plain, in the northwest UK. Built in 1957, the surface has been continually improved so the telescope can operate at shorter wavelengths and keep its place among the leading single-dish telescopes in the world. With the elliptical telescope (right), the Lovell Telescope can be linked to other radio dishes scattered over much of England and Wales, to form part of the Multi-Element Radio-Linked Interferometer Network (MERLIN).*

young engineer called Karl Jansky was sent out to New Jersey to investigate. His big rotating radio aerial could pin down the direction from which the static was coming. After a year, he could distinguish hiss from local thunderstorms, distant thunderstorms, and background static which seemed to come from space. We now know that Jansky had detected radio waves generated out in the regions between the stars of our Galaxy by electrons travelling through the extended magnetic fields of the Milky Way.

Radio astronomers have been depicted by cartoonists as 'listening in' to the crackle and hiss picked up by radio telescopes. Although the early pioneers like Jansky did listen to the static, radio astronomers no longer regard their radio telescopes as 'ears', but more as 'eyes'. They pick up radio waves in much the same way as we all do at home – a radio telescope is really just a very sensitive version of a domestic radio set. Radio waves reach our radios from a whole variety of radio stations, in many different directions, all broadcasting at different wavelengths. We tune the receiver to a particular wavelength, and the radio set extracts the superimposed 'message' of voices or music from the radio waves which carry it from the transmitter to our aerial.

The distant natural radio sources in the Universe, however, simply emit a cacophony of hiss and crackle; 'listening in' tells us little about *where* the individual radio sources are in the sky, or what their size and shape are. For this, a 'radio picture' is needed, and that is what a modern radio telescope provides. If we operated a radio telescope at the wavelengths used for terrestrial broadcasting, the result would not be the latest news bulletin, but a 'photograph' showing the aerials of the radio transmitters.

In practice, radio astronomers must take care *not* to observe at broadcasting wavelengths, because the signals

from artificial transmitters would swamp the faint radio waves from the depths of the Universe, just as the brilliance of the Sun makes optical astronomy impossible during daytime. International conventions have allocated specific wavelengths to radio astronomy (Fig. 7.1). No one is permitted to broadcast at any of these two dozen or so wavelengths, so radio telescopes tuned to them can observe the Universe without the foreground 'glare' of radio transmitters on Earth.

As well as these artificial limitations on radio astronomy, the Earth's atmosphere places restrictions on the wavelengths radio telescopes can use. In theory, radio waves include all radiation of wavelength longer than about one millimetre. The longest wavelengths, however, are reflected by the Earth's upper atmosphere, the *ionosphere*. This was a great advantage to early radio broadcasting pioneers, because artificial transmissions could be bounced around the world by the ionosphere. But long wavelength radiation coming from space is also reflected back out into space. So radio astronomers on the ground cannot pick up radio waves with wavelengths much greater than thirty metres.

At the other end of the scale, radio waves merge into the far infrared regions with a rather arbitrary boundary at one millimetre. Water vapour in the Earth's lower atmosphere absorbs the shortest radio waves. From a radio observatory at sea-level, it is difficult to pick up cosmic radiation of wavelength less than one or two centimetres.

Radio waves with wavelengths in the wide range from two centimetres to thirty metres can penetrate both the ionosphere and lower atmosphere, but the air around is, in fact, surprisingly opaque to most electromagnetic radiation. The radio wavelengths are the only natural 'window' apart from the optical, which allows visible light through, so

Fig. 7.3 *The Cambridge Four Hectare telescope (foreground) consists of thousands of simple aerials (dipoles) strung between wooden posts. Pulsars were discovered by the original half of this very sensitive array. In the background is a dish of the One Mile Telescope, which consists of three dishes in a 1.6-kilometre line.*

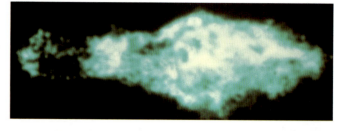

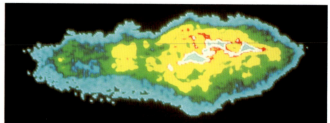

Figs. 7.4–7.5 *Two ways of displaying the same radio view of the supernova remnant 3C 58. Fig. 7.4 shows the radio intensity converted into brightness, similar to an optical photograph. In Fig. 7.5, the intensity is colour coded. The faintest regions are pale blue, and brighter parts dark blue, green, yellow, red, white and pale blue (again) for the intense central regions.*

radio astronomy is the only branch of invisible astronomy whose telescopes can be situated, to our convenience, at sea level. As a result, radio observatories have been built around the world to pick up the natural cosmic 'broadcasts' – in the deserts of New Mexico, the fens of Cambridgeshire, the Caribbean and the Australian outback.

The main components of a radio telescope parallel those of a domestic radio set. At the front there is an aerial or antenna. The energy of radio waves passing the antenna is converted into electrical signals. These are very weak voltage fluctuations, and they are boosted by the second component, the amplifier – in practice, a series of amplifiers which increase the strength of the incoming signal about a thousand million million times. Finally, there is a computer to store the output, and display the signal in some convenient form.

Although we usually think of radio telescopes as large saucer-shaped dishes – like the giant dishes at Jodrell Bank (Fig. 7.2) or Effelsberg in Germany – they can actually be of several different forms. A single wire will act as an antenna to pick up radio waves from space, but its signal will naturally be weak. Some radio telescopes consist of thousands of simple wire *dipole* aerials next to one another, so that the radio waves can be collected over a much larger area and their combined output is considerably stronger. British radio astronomer Tony Hewish constructed one such array near Cambridge in the 1960s; it had 2048 individual wire dipoles spread over a two-hectare field (Fig. 7.3). This 'telescope' was so sensitive that Hewish could look at rapid fluctuations ('twinkling') in the radio emissions from sources in the sky. His student, Jocelyn Bell, noticed that some of the 'twinkling' was actually arriving in the form of regular pulses of radiation, at intervals of a second or so. A month's study convinced the Cambridge astronomer that these *pulsars* were not intelligent broadcasts from an alien civilisation. They are the natural radio emission from the tiny, collapsed cores of old stars.

Such a star is only about 20 kilometres in diameter, but contains rather more matter than the Sun; it is composed of subatomic particles called neutrons, packed so tightly

together that a pinhead of matter from one of these *neutron stars* would weigh a million tonnes! The radio waves are emitted in beams from opposite sides of the star. As the neutron star spins round with a period of about a second, one of the beams regularly sweeps past the Earth, and a radio telescope picks up a 'pulse' of radiation. In most pulsars, the second beam points in such a direction that it misses the Earth, but in some cases its radiation is detected as a weak 'interpulse' between the main pulses.

Unusual telescopes like Hewish's are the exception. Although they are cheap, they are fixed, pointing more-or-less straight upwards. A dish-shaped telescope is much more versatile. It operates very like an optical reflecting telescope: radio waves from space hitting the inside of the bowl are reflected up to a focus, and the telescope can be swung round to look at any part of the sky. But a radio telescope cannot 'photograph' directly the image at its focus. A dipole antenna at the focus picks up and measures the strength of the radio waves, which come from just a small region of sky directly in front of the telescope dish. To build up a 'picture' of the source, the telescope scans back and forth across it, so that its brightness can be measured at every point. The telescope's computer stores the measurements, and builds up a 'map' of the source. This is a two-dimensional array of numbers, the number stored at each point indicating the brightness of the corresponding region of the sky. The map can then be displayed in various ways.

The most direct method is to display the information on a computer monitor screen, with the brightness of each point representing the number stored in the computer – hence showing the brightness of the radio sky. This method produces 'radio photographs' which show realistically how the sky would look through 'radio eyes' (Fig. 7.4). The numbers stored in the computer are equivalent to the array of numbers that represent an optical picture from a CCD electronic detector (Chapter 3). So the techniques of false-colouring and other computer-generated representations can also be used on radio maps (Fig. 7.5).

Such images can be produced by any radio telescope, but

Figs. 7.6–7.9 *Among the world's largest fully steerable radio telescopes is the 100-metre diameter dish at Effelsberg, near Bonn in Germany (**Fig. 7.6**, top left). The outer part of the dish is made of mesh, to reduce wind pressure which distorts the telescope's shape. The valley location shields it from radio interference. The control building has a grandstand view of the telescope (**Fig. 7.9**, bottom right). A typical radio map, of the galaxy IC 342 (**Fig. 7.7**, top right), is colour coded for intensity; the background sky is dark blue, brighter regions are green, yellow, red, then blue, yellow and red again. **Fig. 7.8** (bottom left) shows similar coding applied to a photograph of the Effelsberg telescope itself, with the most brilliant parts coded white.*

Fig. 7.10 *The Arecibo telescope is the world's largest radio dish, 305 metres across. Fitted into in a natural hollow in the limestone hills of Puerto Rico, the curved bowl reflects radio waves to an antenna suspended on cables a dizzying 130 metres above. Although the dish cannot tilt, astronomers can move the antenna to allow the telescope to pick up radiation from different parts of the sky. Observing the longer radio waves, this telescope studies the Earth's ionosphere, pulsars and galaxies – and has been used to search for signals from intelligent aliens.*

Caribbean island of Puerto Rico. The Arecibo dish (Fig. 7.10) is an immense bowl of wire-netting, 305 metres across. The dish reflects radio waves up to a receiving antenna strung on a girder 130 metres above it – the height of a fifty-storey building. The Arecibo telescope's huge area means that it can detect some very faint radio sources – but even this Leviathan can barely resolve detail as fine as the human eye can perceive at visible wavelengths.

However, radio telescopes with such low resolution can perform many useful tasks. The steerable dishes can maintain observation of an individual radio source, timing the pulses from a neutron star, or watching a flare star for one of its unpredictable bursts of radio waves.

Such radio telescopes are also invaluable for surveying the sky. Most stars emit radio waves so faintly that even the most sensitive radio telescopes cannot detect them; the strongest radio sources in the sky are mostly objects that emit very little visible light, and a powerful optical telescope is needed to reveal their appearance at visible wavelengths. So surveys at radio wavelengths are used to locate these sources. A large radio telescope with a resolving power rather poorer than the human eye – a few arcminutes in angular scale – is about right for such work; a telescope with a finer beam would take an inordinately long time to scan the entire sky.

The earliest surveys showed that the brightest object in the radio sky is the band of the Milky Way – as the pioneer Jansky had suggested. The Milky Way outshines even the Sun at radio wavelengths – except when the Sun's surface is broken by the explosion of a solar flare, which shines in radio waves a million times more brightly than the whole of the rest of the Sun. As well as the diffuse band of the Milky Way, the radio sky is dotted with individual, small radio sources (Fig. 12.1). These are not stars. Many of them are so extended that their diffuse, 'woolly' shapes would be seen easily with radio detectors as sharp as the human eye. Some are the rings of *supernova remnants*, the exploding gas shells produced at the death of a star as a supernova; others are the glowing gases of nebulae – like the Orion Nebula – which surround relatively young stars.

Many of the brightest sources actually lie millions of light years away, well outside our own Galaxy. There are titanic explosions at the centres of some distant galaxies, with compact radio sources called *quasars* marking the site of the explosion in the most violent of these galaxies. The emissions from *radio galaxies* mark the aftermath: the energy blown out in the central explosion creates huge radio-emitting lobes on either side of the galaxy itself (Fig. 7.11). Radio detectors, blind to the millions of stars making up the galaxy, see a dumb-bell in the sky, sometimes stretching out millions of light years on either side of the galaxy which has spawned the colossal clouds.

Early radio astronomers named individual sources after the constellations they appear to lie in. The strongest source

astronomers generally want to build the biggest radio telescope they possibly can. Their reasons are similar to the arguments of 'light grasp' and 'resolution' which have driven optical astronomers to build larger telescopes. A bigger dish can collect radio waves over a larger area, and resolve more detail, responding to the small scale structure rather than blurring it out.

The question of resolution is one of the radio astronomer's main headaches. The finest detail which any telescope can reveal depends on the diameter of the telescope mirror (lens, or dish) *relative to the wavelength* it is detecting. Radio waves are almost a million times longer than light radiation, so a radio telescope must be roughly a million times bigger than an optical telescope if it is to resolve the same kind of detail in the sky.

The increase in scale is quite staggering. It means that when the 100-metre diameter Effelsberg radio telescope (Figs. 7.6–7.9), observes at a typical wavelength of 11 centimetres, it actually 'sees' a more blurred view of the sky than does the human eye. An effort to build bigger telescopes encounters a problem with the engineering: how to keep the enormous bowl true to shape as it is tilted from the horizontal to the vertical and buffeted by winds. The dish at Effelsberg, near Bonn, is near the size limit for a fully steerable dish. A new telescope at Greenbank, West Virginia, will be marginally larger, with an oval dish 100 x 110 metres in size.

Astronomers from Cornell University, New York State, accepting that a larger dish can be built but not tilted, have constructed the world's largest radio dish in a natural hollow in the limestone hills near Arecibo, on the

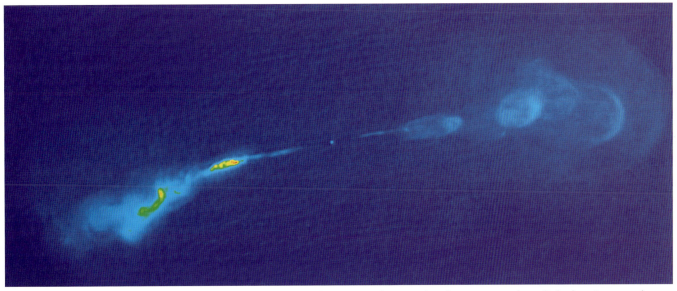

Fig. 7.11 *Hercules A was one of the first radio sources to be discovered: its name indicates early radio astronomers knew little about it, except the constellation it lay in! Hercules A turned out to coincide with a distant galaxy, emitting radio waves a million times more powerfully than our Milky Way. The latest radio image is here colour coded so that bright parts are red, and fainter parts yellow, green and blue. The radio emission comes from two magnetic clouds, a million years across, supplied with energy by jets of electrons which emerge from the galaxy's active core. Each of the bubbles blown by the right-hand jet is bigger than our Milky Way Galaxy.*

in the constellation Cygnus was called Cygnus A (Figs. 12.10–12.12), and so on through the alphabet. Later surveys at observatories around the world have listed too many sources for this simple system to be used. Instead, each observatory has produced its own catalogue – so the same source can appear under different catalogue numbers. The first major catalogue of the northern skies, the Third Cambridge Survey, is the most widely quoted. The nearest quasar, for example, is known as 3C 273 (Figs. 12.30–12.36): it was originally entered as the two hundred and seventy-third radio source in the catalogue.

Radio astronomers have also scanned the sky at different wavelengths, to pick out different kinds of sources. The hot gas in nebulae, for example, shines much more brightly at the shorter wavelengths, as do the cores of the distant quasars. A survey at short wavelengths picks out objects such as these as the brightest in the sky. Most other kinds of radio source consist of high-speed electrons travelling through magnetic fields. They produce radio waves of a type called *synchrotron radiation* which is strongest at the longer wavelengths. A long-wavelength survey is dominated by the synchrotron radiation from the supernova remnants and the distant magnetic lobes of the radio galaxies.

By looking at a radio source's relative strength at different wavelengths, something can be said about what is producing its radiation: whether it is a cloud of hot gas, or a magnetic 'bag' full of electrons. But to find out exactly what is going on, a higher resolution detector is needed to probe the finer details of the source.

Since a telescope's resolution depends both on its diameter and the wavelength that it is observing, there are two ways to improve your view of the fine structure in radio sources: either build a bigger telescope, or use existing telescopes at a shorter wavelength.

Radio astronomers have invented ingenious techniques to make radio telescopes that are effectively several kilometres in size – far larger than the size at which a single dish can be

built. The most successful method – *Earth-rotation synthesis* – was pioneered in the 1960s by the Cambridge radio astronomer Martin Ryle. It relies quite simply on two small radio telescopes, a computer, and the fact that the Earth rotates on its axis. It uses these ingredients to build up what is in effect a huge radio telescope mirror.

Any mirror forms an image at the focus by the merging of radiation reflected from each part of its surface, in such a way that the waves of radiation 'interfere' with one another – the 'crest' of one wave can reinforce the crest of another, or be 'damped' by another wave's 'trough'. In principle, it should be possible to create the effect of a very large optical telescope mirror with just two small mirrors. Keeping one mirror fixed in the centre, the other one can be moved to successive positions around it until it has covered the total area of an imaginary large mirror which is in effect being synthesised. When the mobile mirror is at each position, the image at the focus can be recorded; and eventually all these images can be combined to make the image which the larger imaginary mirror would have seen.

That is the theory; unfortunately it is difficult to make it work for optical telescopes. You have to keep track of the phases of each image (the parts of the waveform at each point in the mirror) in order to add them all correctly at the end; and the vibrations making up light are so rapid that this is an immensely challenging task. Radio astronomy has the advantage here: the phase of the incoming radio wave is preserved in the electrical signal that forms the radio telescope's output, and it is at a sufficiently low frequency that electronic circuits can record it. In addition, the two radio telescopes do not actually have to reflect radiation to a distant mutual focus. This too can be achieved electronically. Each small dish collects radio waves in the normal way, and their electrical outputs are merged to mimic the merging of radio radiation at the focus.

So a large radio telescope *can* be synthesised by two small dishes and some electronic wizardry. There is obviously a practical problem in moving a radio telescope dish

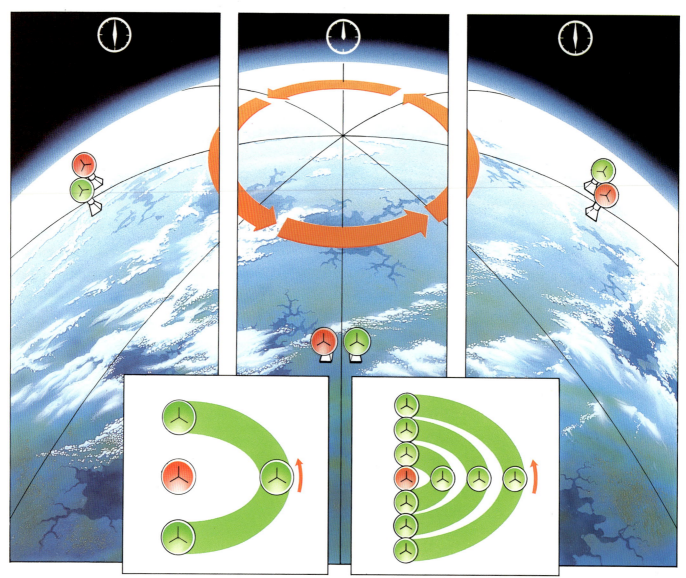

Fig. 7.12 *Earth-rotation synthesis uses a pair of small dishes to mimic a large one. The Earth's rotation carries one telescope around the other, through 180° in 12 hours (left inset). By changing their separation daily, radio astronomers build up a complete synthesised large dish (right inset) – the missing half is added by computer (it contains no new information).*

continually from place to place to cover an area the size of the large imaginary dish, but for this we can make use of the Earth's spin (Fig. 7.12). Imagine looking down from a radio source which happens to lie above the Earth's North Pole. A radio astronomer in middle-latitudes – say England, the United States or Australia – has set up two small radio dishes on an East-West line. The Earth rotates anticlockwise and the dishes travel with it. As they do so, their *orientation relative to each other in space* changes too. We see this effect from our position above the North Pole. Relative to one dish, the other circles it in an anticlockwise direction; it moves round continuously, until after twelve hours this dish seems to have moved to the other side of the first dish. If we regard the first dish as the centre of a synthesised mirror, the second has travelled in a semicircle around it, to trace out half a ring of mirror surface, some distance out from the centre.

The radio astronomer then moves the second mirror slightly closer to the first, and lets the Earth's rotation create another, slightly smaller semicircular ring of mirror. On moving it nearer for another twelve-hour run, a still

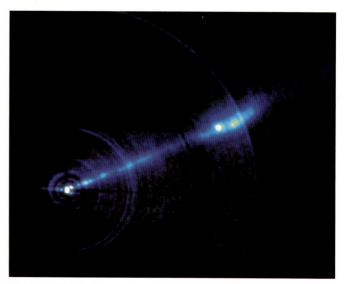

Fig. 7.13 *The Earth-rotation technique has caused the faint false rings in this detailed view of the jet of electrons from the galaxy NGC 6251, obtained by the Ryle Telescope near Cambridge, England.*

Fig. 7.14 *Five dishes of the Australia Telescope form the first array to study the far southern part of the sky, which includes our nearest neighbour galaxy, the Large Magellanic Cloud. These dishes at Narrabri, in northwestern New South Wales, can be linked to another dish near the Siding Spring optical observatory and to the large Parkes Telescope further south to form an array 300 kilometres across. Eventually, they may be linked to radio dishes around the continent to form an array as large as Australia itself.*

smaller ring of the mirror is 'filled in'. Eventually, an entire half-mirror has been synthesised. Theory shows that the information from the 'missing' half of the mirror is mathematically related to that obtained from the synthesised half, so the computer can add in the missing information automatically. The technique puts into the computer all the information that would have been gathered by a complete dish equal in size to the largest separation of the small dishes.

This information is stored as a set of electronic intensities and phases, each measurement corresponding to one relative position of the dishes. A mathematical technique called Fourier analysis can be used to convert this mass of figures directly into a view of the radio sky – stored neatly as an array of numbers, just like the output of a scanning single dish. The Earth-rotation synthesis telescopes do not even need to scan back and forth. Each dish points constantly at a particular radio source, and the Fourier analysis effectively does the scanning within the computer.

Ryle's first Earth-rotation synthesis telescope was 1.6 kilometres long. It was followed by another 5 kilometres long, now called the Ryle Telescope. It has eight dishes, so sixteen rings of the mirror can be synthesised simultaneously. Similar telescopes have been built at Westerbork in the Netherlands and in Australia (Fig. 7.14).

These telescopes revolutionised our knowledge of radio sources, particularly of radio galaxies and quasars. Their fine resolution of detail showed why a galaxy's central explosion can produce the distended radio-emitting lobes. They revealed narrow glowing 'jets' of matter (Fig. 7.13) stretching out from either side of the core, leading to the lobes. These jets are evidently beams of fast electrons, shot out at high speed from the core like the electron beam from the back of a television tube. Where these beams collide with the tenuous gas outside the galaxy, they are reflected back in disorder, like the jet of water from a garden hose splashing back from a wall. The electrons' motion constitutes a disorderly electric current which generates a tangled magnetic field in these regions. As the high-speed electrons move through this self-generated field, they emit synchrotron radiation at radio wavelengths.

The first Earth-rotation synthesis telescopes had problems looking at radio sources far from the North Pole of the sky. As seen from a radio source above the Earth's North Pole – in the direction of the celestial North Pole – the dishes in these arrays trace out semicircular rings as the Earth rotates. But the view from a radio source away from the pole is foreshortened; each dish traces out an oval path relative to the others. The synthesised dish is thus not a circular mirror, but an oval one. This shape resolves details of a radio source less well in the North-South direction.

The answer is to include more dishes in the radio telescope array, in lines that extend some distance North-South, rather than exactly East-West. As seen from the

radio source, each dish now traces out a much more complicated arc relative to the other: the synthesised mirror is made from many differently shaped arcs of mirror surface. The telescope computer can, however, calculate the Fourier transform from the measurements made along these arcs, to produce the appearance of the radio sky as reflected in this large 'partial mirror'. Just as an optical mirror with pieces missing cannot form an accurate image, so the partial mirror of the radio telescope produces distortions in the image it makes. But the distortions are now routinely removed using computer programs designed specifically to 'clean up' maps of the radio sky.

The most ambitious of these radio telescopes is the Very Large Array (Fig. 7.15) sited in the desert near Socorro in New Mexico. The VLA has 27 dish aerials each 25 metres across, which is large by the standards of a radio telescope anyway. They can move to various positions along the arms of a Y-shaped railway network. The whole array effectively makes a radio dish 27 kilometres in diameter. When the VLA is observing at its shortest wavelength of 1.3 centimetres, it can resolve details 0.13 arcseconds in scale – almost a thousand times better than the human eye, and finer than any Earth-based optical telescope can see.

Arrays like the Australia Telescope and the VLA have their dishes connected electronically by buried cables or waveguide tubes to the central control building. The VLA represents about the largest size possible for directly linked telescopes, but a larger array can be built if instead the signals from each dish are amplified, then fed into an ordinary radio communications transmitter which beams the signal to the control centre. Astronomers at Jodrell Bank have connected several isolated dishes in England by radio links, to make a dish effectively 230 kilometres across. This network is called MERLIN – the Multi-Element Radio-Linked Interferometer Network. It can perceive details as small as 0.01 arcseconds in size when it observes at 1.3-centimetre wavelength. In practice, though, longer wavelengths are often used which give MERLIN about the same resolution as the VLA. As a result, the variation in the structure of a radio source at different wavelengths can be investigated.

The size of radio telescopes can, however, be pushed even further to mimic the performance of a radio telescope the size of the Earth. In the technique of Very Long Baseline Interferometry (VLBI), astronomers at widely separated radio observatories look at the same radio source

Fig. 7.15 *A desert rainbow forms a spectacular backdrop to the most sophisticated purpose-built radio telescope array, the Very Large Array in New Mexico. The dishes are seen here unusually close together. They can be spaced out on railway tracks 21 kilometres long to synthesise a correspondingly large telescope.*

simultaneously, and record the output of their telescopes, along with the signals from an atomic clock which keeps an extremely accurate track of time. The tapes are flown to a common centre, where they are played back, with the atomic time signals keeping them exactly synchronised. The outputs from the telescopes can then be added together electronically just as if the telescopes had actually been connected while they were observing.

With VLBI, details 0.001 arcseconds in size can be resolved – a thousandth the scale of the smallest detail an optical telescope can see, and equivalent to making out a pinhead at a distance of 200 kilometres. But there is a price to pay. Two radio telescopes separated by almost the diameter of the world form only two very tiny portions of the mirror they are trying to synthesise. In fact, they cannot form a proper image at all, although they can detect whether or not a radio source has very fine details within it. Additional radio telescopes need to be involved to synthesise a more complete VLBI mirror. The combination of the results from, typically, half a dozen telescopes can produce fine-scale images – the most detailed of all the images formed at any wavelength.

VLBI with several telescopes involves complex international deals to make sure the individual dishes are all available at the same time. To solve this problem, American radio astronomers have built a network of radio telescopes especially for VLBI. The Very Long Baseline Array (VLBA) consists of ten dishes similar to those making up the VLA, but spread over a region 8000 kilometres across; from Hawaii to the US Virgin Islands, and from Texas to Massachusetts. It will regularly make detailed maps of the smallest radio sources, and should easily show the gas circling massive black holes in the centres of radio galaxies and quasars.

The nearest quasar, 3C 273, for example, is a powerful explosion at the core of a galaxy lying some 2000 million light years away. It is a compact blaze of light and other radiations, less than a light year in size. When we look at the galaxy with telescopes that are sensitive to radiations at wavelengths from X-ray to infrared, however, our telescopes or the Earth's atmosphere blur it out to a fuzzy blob which obscures all detail within 5000 light years of the quasar's centre. But a radio VLBI network can probe a thousand times smaller to reveal that the core is ejecting a jet of matter only a few dozen light years long (Fig. 12.36).

The quasars and their elderly relatives the radio galaxies, together with the supernova remnants and the hot gas clouds in our Galaxy, all radiate over the whole range of radio waves, and so can be observed at any chosen wavelength. But there is also a different kind of radio emitter in the sky which radiates at just one or a few specific wavelengths. These are atoms or molecules of gas, in the almost empty space between the stars of a galaxy. To detect them, radio telescopes must be 'tuned-in' precisely to the relevant wavelength.

8 milky way system

THE LUMINOUS BAND of the Milky Way stretching across the night sky is a broadside view of the galaxy in which we live. Astronomers call it the Milky Way Galaxy, or simply the Galaxy. It is a vast swarm of stars, gas and dust shaped like a Catherine-wheel with a bulging centre, and measuring some 100 000 light years from edge to edge. Its main constituent is stars: about 200 000 million in total. Our Sun is just one very ordinary star amongst this unimaginable number. Space between the stars is filled with interstellar matter: gases, dust grains, magnetic fields and very fast subatomic particles called cosmic rays. There is sufficient gas and dust to build a further 20 000 million stars.

From the outside, our Galaxy would look rather like the Andromeda Galaxy (Fig. 10.2); but we see the Galaxy from within, and as a result it is all around us. To 'see' our Galaxy in its entirety, we must look at the whole sky. Most of the images on the next few pages are thus views of the entire sky, spread out into an oval frame in much the same way that the surface of the Earth can be represented by a single oval map. Seen edge-on, the flat disc of the Galaxy appears as the narrow band of the Milky Way. The sky is 'unfolded' such that the left-hand end of the Milky Way on the map in reality is joined up with the right-hand end and the centre of each map is the view directly towards the centre of the Galaxy. Other images just home in on the narrow band of the Milky Way.

It is impossible to photograph the entire sky at one time, so astronomers at the Lund Observatory have prepared this optical view (**Fig. 8.1**) by redrawing details from a mosaic of photographs onto a single drawing. It shows the positions and brightnesses of all the stars visible to the unaided eye, a total of about 7000 individual images.

The Milky Way band consists of thousands of millions of more distant stars, too faint to be seen individually. In the plane of the Galaxy's disc (along the 'equator' in these projections) there are plenty of distant stars and the sky appears bright. But in other directions, we are looking 'up' or 'down' through the narrow extent of the Galaxy's thickness; there are few distant stars and the sky is dark. Thus we get the appearance of a bright band round the sky.

Although this projection distorts the familiar constellations, some patterns can

8.1 *Optical, photographic mosaic, Lund Observatory, Sweden*

be made out. On the extreme right, just below the Milky Way, is Orion, with the brightest star, Sirius, just above. The Plough is at the top left, upside down. Vega is the bright star above the Milky Way left of centre, and below is a prominent dust band which appears to split the Milky Way in the constellation Cygnus. Almost mirroring this on the right is the small dark triangle of the Coal Sack. Three small bright 'clouds' below the Milky Way are neighbouring galaxies, the only objects here which are not part of our Galaxy. On the left is the small oval of the Andromeda Galaxy. In the right half lie the two satellite galaxies of the Milky Way, the Small (left) and Large (right) Magellanic Clouds.

Although these are separate galaxies, they are attached to ours not only by gravity but by streamers of invisible hydrogen gas.

The dark patches within the Milky Way are dense clouds whose dust particles are blocking off the light from stars beyond. In fact, there are many more stars towards the Galaxy's centre, and the optical view should show a really brilliant region at the centre of this picture, but where there are more stars, there is also more dust. This absorbs much of the extra light, to make the Milky Way rather uniform.

But other wavelengths can cut right through the obscuring dust. One of the major triumphs of the new astronomy has been to lay bare the far reaches of our

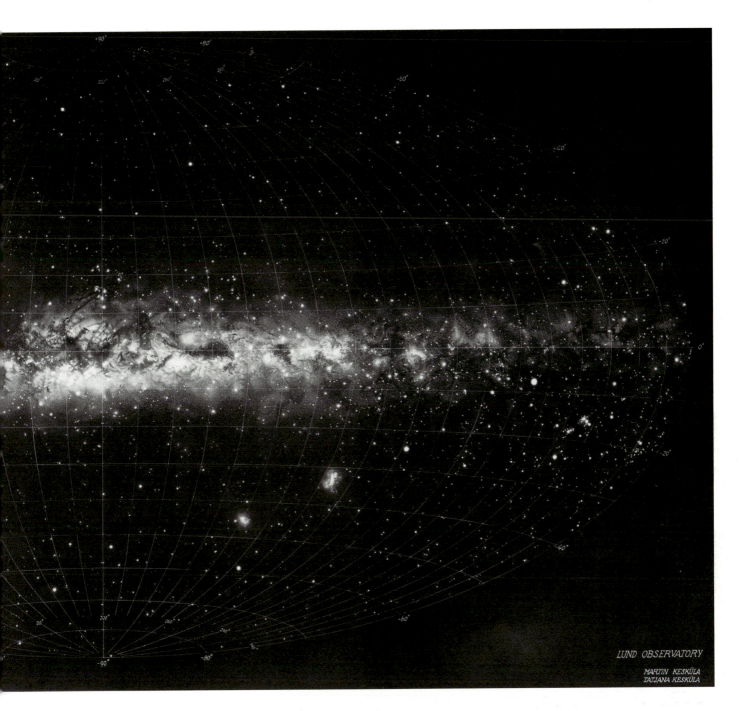

Milky Way Galaxy. **Fig. 8.2** is an infrared view looking towards the centre of our Galaxy and covering 180° from side to side – corresponding to the central half of Fig. 8.1.

In Fig. 8.2, we pick up infrared light from even the most distant reaches of our Galaxy. The yellowish horizontal stripe is the radiation from ordinary stars lying in the thin disc of the Milky Way. The thicker white region in the middle reveals the brilliant bulge of older stars lying around the centre of our Galaxy. The odd 'peanut' shape indicates the bulge is not circular in shape, like a hub-cap, but forms an elongated bar that is angled towards the Sun.

8.2 *Infrared, 1.2 μm (blue), 2.2 μm (green), 3.4 μm (red), Cosmic Background Explorer*

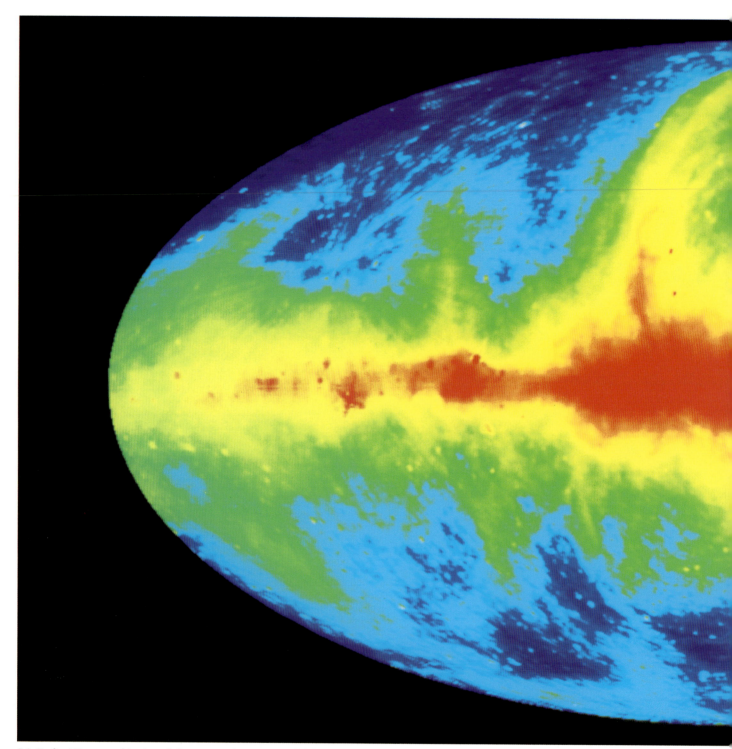

8.3 *Radio, 73 cm, combination of observations from 100 m Effelsberg, 76 m Lovell and 64 m Parkes telescopes*

The radio sky is dominated by the Milky Way, as seen in this whole-sky map (**Fig. 8.3**). It is drawn out in galactic coordinates, in the same way as the optical view (Fig. 8.1) so the two can be compared directly. In this false-colour view, the brightest regions are coded red, and successively fainter regions yellow, green and blue.

No single radio telescope can see the whole sky (unless it is sited on the Equator), and this radio view has been compiled from observations made with three of the world's largest radio telescopes: the 76-metre at Jodrell Bank, England; the 100-metre at Effelsberg, Germany; and the 64-metre at Parkes, in Australia. All the telescopes observed at a wavelength of 73 centimetres, and built up different parts of the map by scanning back and forth across the sky, measuring the radio brightness at each point.

Dozens of bright individual sources stand out in Fig. 8.3. At the centre of the red cross on the left is Cassiopeia A (Figs. 6.22–6.26), a source so brilliant it creates false 'spikes' like a bright star on a photograph. The large red area to its right is 'Cygnus X', a confused mass of radio sources lying one behind another along our local spiral arm. The small source in the upper right of Cygnus X is Cygnus A (Figs. 12.10–12.12), a radio galaxy which by chance lies almost in the plane of the Milky Way.

The small red blob at the extreme right is

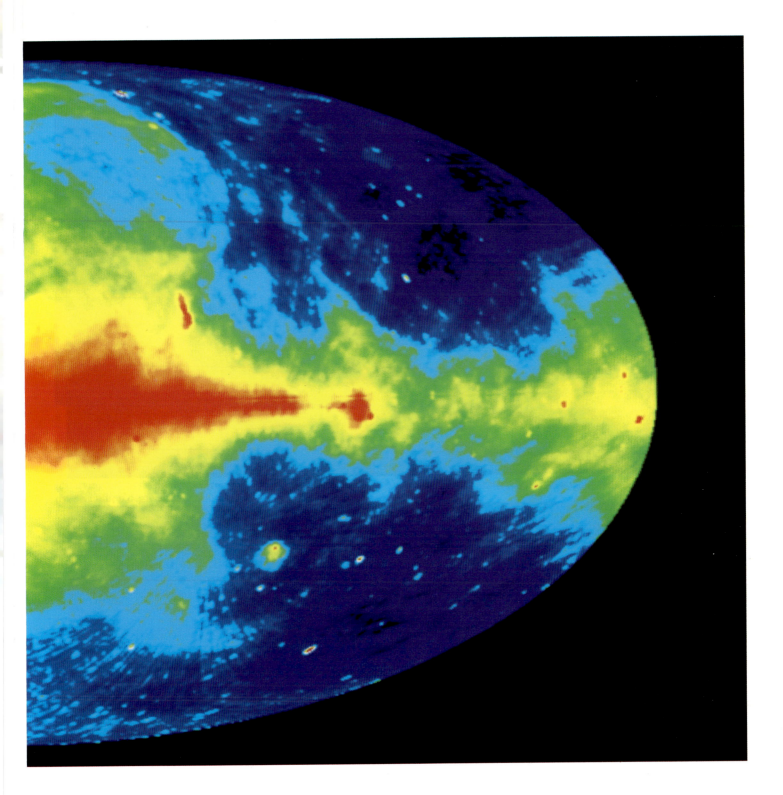

the Crab Nebula (Figs. 6.9–6.18). The larger red patch half-way in from the right is the Vela region, where the Vela supernova remnant (Figs. 6.27–6.33) is superimposed on the emission from our local spiral arm, heading in the opposite direction from Cygnus. To its left, and well below the Milky Way, an extended radio source marks out our neighbour galaxy, the Large Magellanic Cloud (Figs. 8.19–8.23). The elongated source above the Milky Way (a red vertical 'bowtie' on a yellow background) is the nearest radio galaxy, Centaurus A (Figs. 12.2–12.9).

The radio emission from the Milky Way itself is mainly synchrotron radiation, generated in interstellar space by fast electrons moving through the Galaxy's magnetic field. The radio map is not obscured by dust (unlike the optical view, Fig. 8.1), and the central parts of the Galaxy appear in their true brilliance.

Nearer to us, great filaments marking the magnetic field stretch out of the Galaxy's disc. The most obvious filament, the North Polar Spur, stretches from near the centre of this view upwards, almost to the top of the map. Here it bends over and comes back down (near the position of Centaurus A) to make a complete loop. It is a giant bubble of gas, blown by hot stars in Scorpius. These stars are visible in Fig. 8.1, above the Milky Way just to the right of centre.

Infrared and radio astronomers are well placed to investigate what lies at the heart of the Milky Way. These radiations can slice right through the intervening dust.

Fig. 8.15 is an infrared view of the central 30 light years of the Galaxy, at a magnification 500 times greater than the optical photograph (Fig. 8.11). It is colour coded according to wavelength. The shorter wavelengths in this composite (blue images) show only stars near to the Sun, because this radiation suffers significant absorption as it passes through the densest dust clouds lying between us and the centre of the Galaxy. But stars at the galactic centre shine through at the longer wavelengths, and appear orange and red here.

This infrared image shows clearly how stars become ever more closely packed towards the centre of the Galaxy. In the middle of Fig. 8.15 is the dense star cluster at the Galaxy's heart, containing some 10 million stars packed into a region only 10 light years across. If we lived here, the nearest stars would be only a light week away, and hundreds of stars would appear brighter than the Full Moon.

The stars in the central five light years are shown in even greater detail in Fig. 8.16, another infrared image colour coded for wavelength. Most of these objects lie at the galactic centre, rather than being foreground stars, so the infrared colours give an idea of their temperatures: blue being hot and red comparatively cool. If this were our part of the Galaxy, only two stars – the Sun and our nearest neighbour the Alpha Centauri system – would appear in this frame. Instead, it is filled with dozens of stars. Each is thousands of times more brilliant than the Sun, yet they are known only by their infrared source (IRS) catalogue numbers.

The brightest star, at the top centre of the frame, is a huge cool supergiant star, IRS 7. To its lower right is another cool star (IRS 3) wrapped in a dense cocoon of dust: only the longest wavelength infrared can struggle out, giving it an orange hue in this representation. In the centre is a group of a dozen or more very energetic stars (IRS 16), appearing blue here. Many of these may be very hot Wolf-Rayet stars, massive old stars that have lost much of their outer layers.

Before these observations, some astronomers had invoked an active galactic nucleus to provide powerful ultraviolet radiation that lights up and heats the gas around the galactic centre – for example, the strand of gas (red) to the lower right in Fig. 8.16. But this infrared image reveals enough Wolf-Rayet stars in the galactic centre to do this job quite adequately. Even worse for those in favour of an active nucleus in our Galaxy, the infrared view shows nothing at all at the exact position (marked by the cross) of the galactic centre.

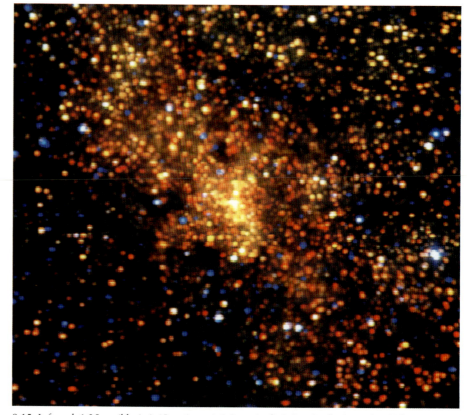

8.15 *Infrared, 1.25μm (blue), 1.65μm (green), 2.2 μm (red), 3.9 m Anglo-Australian Telescope*

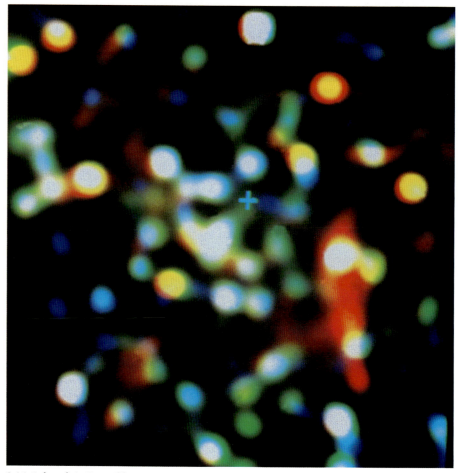

8.16 *Infrared, 1.65 μm (blue), 2.2 μm (green), 3.45 μm (red), 4 m reflector, Cerro Tololo Interamerican Observatory*

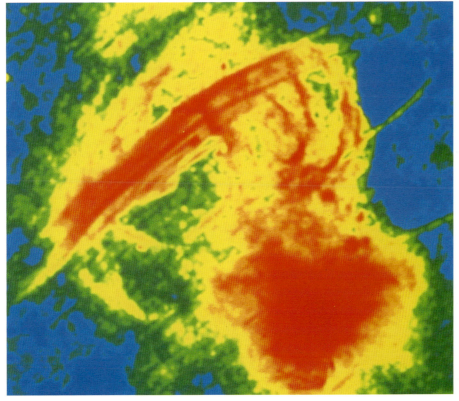

8.17 *Radio, 20 cm, Very Large Array*

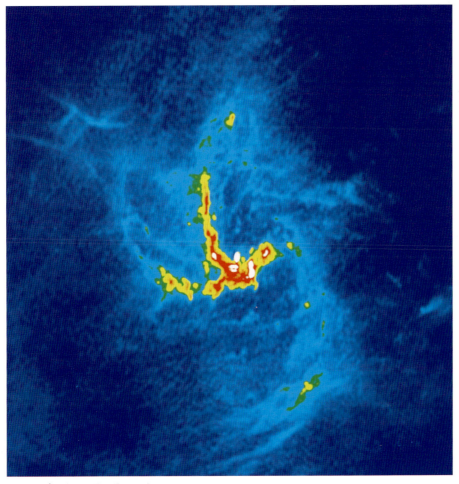

8.18 *Radio, 2 cm, Very Large Array*

But radio observations have always been the ultimate touchstone for determining activity in galactic nuclei. And radio telescopes reveal some pretty odd activity near the centre of the Milky Way.

Fig. 8.17 is a radio view of the galactic centre region, colour coded for intensity. The brightest regions are red, and fainter parts yellow, green and blue. This image is 180 light years across, and includes the centre of the Galaxy in Sagittarius A West, the large red complex to the lower right.

The most striking feature of Fig. 8.17 is the 'Arc': the region of bright radio-emitting filaments at the upper left, running perpendicular to the line of the Milky Way. The individual strands in the Arc are over 100 light years long, yet less than half a light year thick. Magnetic fields are undoubtedly responsible for this filamentary structure, and the ionised gas here must be feeling magnetic forces every bit as strong as the gravitational forces from the massed stars in the region. This is proof of a powerful dynamo in the galactic centre.

A close-up of Sagittarius A West appears in **Fig. 8.18**, where the colour shows radio brightness once more, with successively brighter regions dark blue, pale blue, green, yellow, red and white. This striking spiral of hot gas is only ten light years across: it encloses the bright stars seen in the second infrared image (Fig. 8.16).

Here we find even stronger evidence that our placid Milky Way has an active nucleus at its heart. The gases in this spiral are orbiting the galactic centre so fast that they must be feeling a gravitational tug equivalent to the pull of six million suns. Yet, when we use infrared observations to calculate the number of stars within the spiral, they total only three million suns. The balance must lie in some compact object emitting hardly any radiation: by a process of elimination, it must be a black hole as heavy as three million suns.

And Fig. 8.18 reveals precisely where this hidden monster lurks. Just above the centre of the multicoloured spiral is a single intense source of radio waves (small white oval): Sagittarius A*. This radio source is unique in our Galaxy. First, while everything else in the Milky Way is on the move, as they follow their orbits, Sagittarius A* is absolutely stationary and must therefore lie right at the Galaxy's centre. Astronomers use it as their 'Greenwich Meridian' of the Galaxy, and the fiducial crosses shown in Figs. 8.13 and 8.16 actually show the position of Sagittarius A*. Secondly, it is the only brilliant radio source that is extremely tiny, much smaller than our Solar System. The radiation from this source fits what we would expect from gas circling a black hole of three million solar masses. In all likelihood, Sagittarius A* is a mini-quasar at the heart of the Milky Way.

The Large Magellanic Cloud

OUR GALAXY'S NEAREST neighbour in space is a nondescript galaxy called the Large Magellanic Cloud. Although it is relatively small and dim, the Large Magellanic Cloud is so close to us on the cosmic scale that it can be seen easily with the naked eye. This 'cloud' and its diminutive companion in the southern skies, the Small Magellanic Cloud, are named after the Portuguese navigator Ferdinand Magellan who first reported them in 1521 during his epic circumnavigation of the Earth.

The Large Magellanic Cloud lies 169 000 light years from us, and its companion further off at 200 000 light years – roughly twice the diameter of the Milky Way Galaxy. Even so, both Clouds lie at less than one-tenth the distance of the nearest major galaxy, the great Andromeda Galaxy (Fig. 10.2). The Magellanic Clouds are presently heading away from our Galaxy, having passed closest to it about 500 million years ago. They are satellites of the Milky Way, orbiting it every 6000 million years just as the Earth orbits the Sun – making the Clouds a part of the 'Milky Way system'. Eventually, they will be torn apart and will merge into our Galaxy.

The Large Magellanic Cloud consists of some 10 000 million stars (one-twentieth as many as the Milky Way) and enough gas to make one new star for every ten in the galaxy at present. It is a microcosm of our own Galaxy, and in many ways the Large Cloud is easier to study because we can see immediately the locations of stars and gas clouds relative to one another.

The photograph of the Large Magellanic Cloud (**Fig. 8.19**) is a composite of three black-and-white pictures taken through blue, green and red filters on the large Schmidt telescope at the European Southern Observatory in Chile, and recombined to produce a sharp, true-colour photograph. It covers a region of sky 4° (eight Moon breadths) across, some 12 000 light years, or one quarter the total extent of the galaxy. In the past, the Large Magellanic Cloud has been classified as an 'irregular galaxy', but its central stars clearly concentrate into a broad 'bar' running across this picture, so it is actually a smaller relative of the barred spiral galaxies.

Most of the stars in the bar are yellow, orange and red – old stars which have existed since the Large Magellanic Cloud was born soon after the Big Bang. But groups of white and bluish-white stars show regions where stars have been born more recently. The youngest of all are still lighting the gases they were formed from – nebulae glowing red in the light from hydrogen atoms. The largest gas cloud (top left) is the massive Tarantula Nebula (Figs. 8.24–8.27).

8.19 *Optical, true colour, 1 m Schmidt telescope, European Southern Observatory*

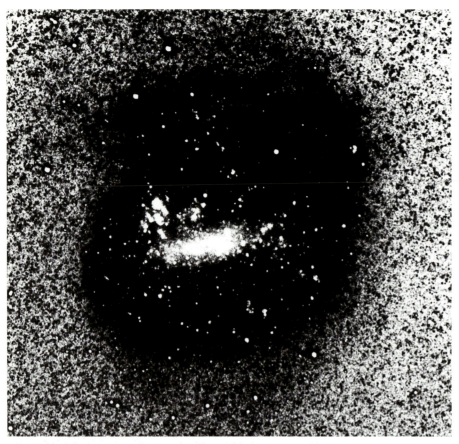

8.20 *Optical, high-contrast photographically-amplified negative and low-contrast positive, Hasselblad SWC camera with 1:4.5/38 mm objective, piggybacked on GPO double astrograph, European Southern Observatory*

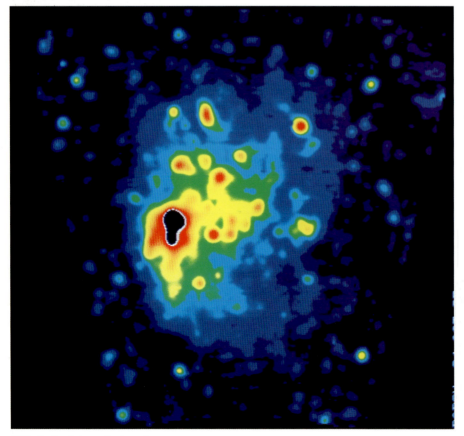

8.21 *Radio, 2.1 cm, Parkes telescope*

There's much more to the Large Magellanic Cloud than meets the eye, as demonstrated by these four images.

You don't even need to move away from optical wavelengths to come up with some surprises. **Fig. 8.20** is an image of the Large Magellanic Cloud at visible wavelengths, showing that the galaxy is much larger than ordinary photographs indicate. This is a composite of two exposures. The positive image reveals the Large Magellanic Cloud as seen in ordinary photographs. It is superimposed – to the same scale – on a negative image made from a photographic plate pushed to the limits of sensitivity. The plate was baked in a mixture of hydrogen and nitrogen, then exposed for 90 minutes: careful printing in the darkroom later enhanced the faintest details on the plate.

The furthest extremities of the Large Magellanic Cloud seen in Fig. 8.20 cover 50 000 light years, making the galaxy 50 per cent larger than previously believed. It covers 15° of the sky. (The dark streak at the bottom left is merely a foreground dusty patch of interstellar 'cirrus' in our Galaxy.) Towards the top, these faint far-flung stars form four or five separate arcs. Some astronomers have interpreted these as rudimentary spiral arms; others as signs that the Large Magellanic Cloud has swallowed up several smaller galaxies in the distant past. Either way, the faint farflung regions of this galaxy contain its oldest inhabitants.

Fig. 8.21, on the other hand, pinpoints regions where stars are being born right now. This radio image is colour coded for brightness, with dim regions blue, brighter parts green, yellow and red, and the most brilliant part (rather confusingly!) black. This image covers the region that appears dark black in Fig. 8.20.

At radio wavelengths the stars making up the bar of the Large Magellanic Cloud just do not register. The emission seen in Fig. 8.21 comes mainly from gas in nebulae that has been heated to 10 000 K by young hot stars in the vicinity – the same process that makes the Orion Nebula a strong radio source (Fig. 4.5).

Many of these nebulae can be discerned in the optical photograph (Fig. 8.19), but they are difficult to pick out against the confusing rash of stars. The radio waves also cut through dust in the nebulae, which obscures some from optical view. The radio image thus gives a much clearer impression of the distribution of star-forming regions across the Large Magellanic Cloud. They are clearly scattered over a much larger region than the galaxy's bar, and with a bit of imagination you can begin to trace out patchy spiral arms in Fig. 8.21.

The radio emission is rather sharply confined at the left-hand side of the Large Magellanic Cloud, where the brightest region in Fig. 8.21 coincides with the great Tarantula Nebula. Recent measurements show that our companion galaxy is moving to the left in the sky, so this bright band of radio emission may represent a cosmic bow-shock: a piling-up of interstellar gas – with subsequent star-formation – where the Large Magellanic Cloud is running into tenuous intergalactic gas. The external gas can flow freely past the stars of the galaxy, so it invades well inside the Large Magellanic Cloud before being stopped by the compressed interstellar gas at the bow-shock.

The hot young stars in the Large Magellanic Cloud give themselves away by emitting ultraviolet light. This radiation is absorbed by the Earth's atmosphere, so ultraviolet views of the cosmos must be obtained by high-flying telescopes – and **Fig. 8.22** was taken with one of the furthest flung of all space telescopes, carried by the Apollo 16 astronauts on their journey to the Moon in 1972. This is how the Large Magellanic Cloud would look to eyes sensitive to radiation of one-quarter the wavelength human eyes can see. It covers roughly the same region of sky as the radio view (Fig. 8.21).

In Fig. 8.22, we are seeing brilliant hot stars, of spectral types O and B, that have shrugged off the dark nebulae in which they were born. Even a trace of dust can hide a star's ultraviolet light. So the dust-laden Tarantula Nebula does not put on a particularly good show here. It is overshadowed by a region to the North (near the top of the image), where a cluster of young stars has blown its natal gas well away into space. It is possible to pick out some of these OB associations in the colour photograph (Fig. 8.19) as loose clusters of bright bluish-white stars.

An X-ray view of the Large Magellanic Cloud reveals more of the life – and death – of these heavyweight stars. **Fig. 8.23** covers just the left half of the galaxy, some 2½° of the sky, where most of the action is. It is colour coded for temperature, from one million K (red), through yellow and green to blue for the very hottest (over six million K).

Young heavyweight stars are violent stars: they emit powerful superhot winds, and die as supernovae which eject yet more very hot gas which eventually diffuses out into space. An extended halo of hot gas (green) is visible in Fig. 8.23 around the starbirth region that contains the Tarantula Nebula.

And the violence continues after their demise. To the right of centre (green and black spot) is a young supernova remnant, N132D. Other supernovae have left neutron stars – and occasionally black holes – that become cosmic powerhouses when they draw matter from a close companion star. The hot object (blue and black) on the left of this image is LMC X-1, a neutron star that is ripping gas from its still-living companion.

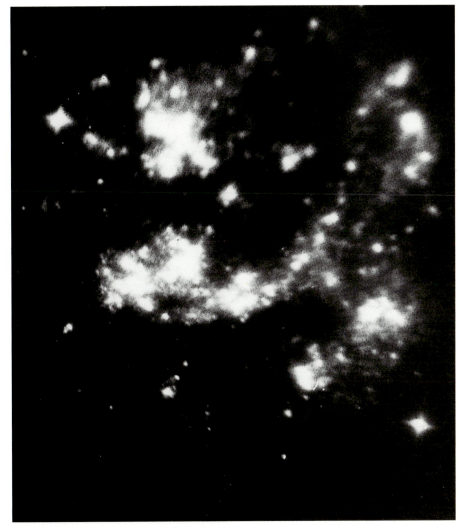

8.22 *Ultraviolet, 125–160 nm, Apollo 16 far UV camera*

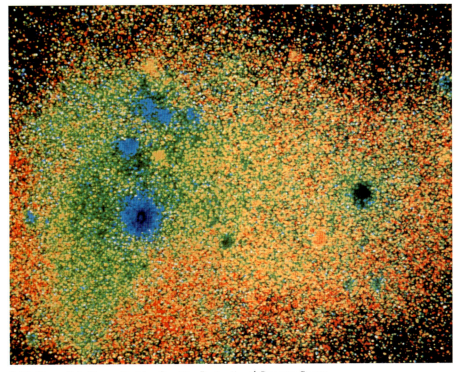

8.23 *X-ray, 0.6–10 nm, Position Sensitive Proportional Counter, Rosat*

9 ultraviolet astronomy

ULTRAVIOLET RADIATION – in small quantities – is beneficial, building up vitamin D within our bodies. But anyone who has fallen asleep under the midday Sun has felt the other effect of ultraviolet radiation: too much exposure kills the living cells of our skin leaving the raw blisters of sunburn.

Our shield against this dangerous radiation is a diffuse layer of invisible ozone gas. Ozone is an unusual form of oxygen. The ordinary oxygen molecules we breathe consist of two atoms joined together, while the ozone molecule is a triplet of oxygen atoms. Ozone gas is transparent to ordinary visible light, but it absorbs ultraviolet radiation very strongly.

The ozone layer has been essential for the existence of life on Earth. Early life evolved in the protection of the oceans, and animal life could only emerge onto dry land because the ozone layer shielded it from the lethal solar ultraviolet. And the opening of an 'ozone hole' above the Antarctic is giving rise to concerns about the effect of extra ultraviolet on life today. But astronomers see the ozone layer as a mixed blessing. As well as blocking solar radiation, the ozone layer absorbs all other ultraviolet rays coming from space. Astronomers trying to see the Universe in the ultraviolet are as badly off as optical astronomers faced with a completely overcast sky; in fact, they are in a worse

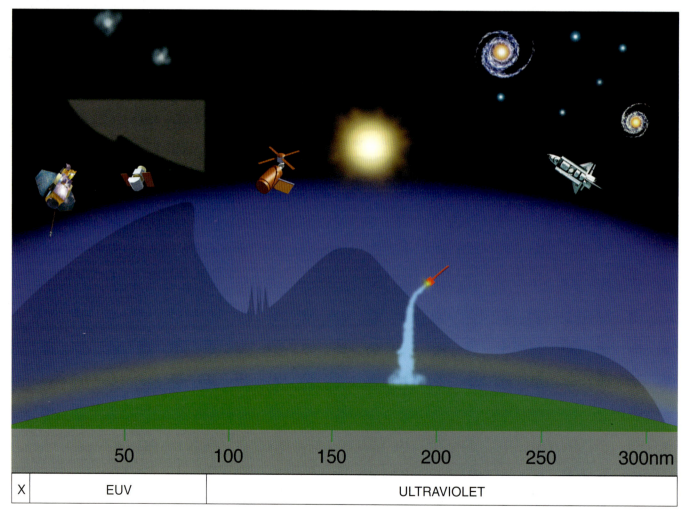

Fig. 9.1 *For astronomers, ultraviolet wavelengths run from the border with X-rays (X) at 10 nanometres (nm) up to the beginning of the optical 'window' at 310 nanometres. The entire band is absorbed by gases in the Earth's atmosphere: at longer wavelengths by the famous ozone layer (shown in yellow) 10 to 50 kilometres above the Earth's surface, and at shorter wavelengths by oxygen and nitrogen at even higher altitudes. Rockets or – preferably – satellites must carry ultraviolet telescopes above these absorbing layers. One regular payload on the space shuttle (right) is the Astro ultraviolet observatory, which has investigated hot stars and the structure of galaxies. The Skylab space station (centre) carried a suite of telescopes for imaging the Sun's hot atmosphere. The Extreme Ultraviolet Explorer (left) and the Wide Field Camera on Rosat (far left) must peer through a 'fog' of interstellar gas (grey wedge at upper left) to investigate the Universe at extreme ultraviolet wavelengths (EUV). Hydrogen in space absorbs wavelengths shorter than 91 nanometres, and helium radiation shortward of 50 nanometres.*

Fig. 9.2 *The last of three Orbiting Astrophysical Observatories undergoing prelaunch tests at Kennedy Space Center, Florida. Renamed Copernicus after launch, the satellite obtained the first high-quality ultraviolet spectra with the largest reflecting telescope then launched. Copernicus lasted nine years before it was switched off in 1981.*

situation, because the 'ozone clouds' never clear away. The ozone layer forms a continuous, permanent, opaque cover to the sky.

Strictly speaking, the ultraviolet region of the spectrum (Fig. 9.1) consists of wavelengths shorter than 390 nanometres – the wavelength of violet light, the shortest radiation visible to the human eye. The ozone layer is, however, transparent to the longer ultraviolet radiation, down to a wavelength of 310 nanometres. So the longer ultraviolet (310 to 390 nanometres wavelength) can penetrate down to sea-level. (It is this radiation which gives us a suntan.) The sky in the longer ultraviolet can be studied in much the same way as the visible light from the Universe, using ordinary optical telescopes with the same kind of electronic detectors.

From the astronomer's point of view, it is more logical to draw the distinction at the wavelength where the ozone layer limits observations. *Ultraviolet astronomy*, therefore, is the study of wavelengths shorter than 310 nanometres, and it involves a whole new technology because ultraviolet telescopes must be raised above the ozone layer.

Astronomers observing the long wavelength infrared radiation can overcome the problem of atmospheric absorption by putting their telescopes on high mountains (Chapter 5). Ultraviolet astronomers have a much worse problem, since the ozone layer lying way up in the stratosphere is three times higher than the summit of Mount Everest. The first unmanned ultraviolet telescopes were flown on high-altitude balloons, like the weather balloons which meteorologists use to investigate the stratosphere. Later telescopes were launched on brief rocket flights which lifted them above the ozone layer for a few minutes at a time. The best answer of all is an ultraviolet satellite. From an orbit above the Earth's atmosphere, a satellite can survey the ultraviolet sky for years on end.

The earliest ultraviolet satellites, for example the second Orbiting Astronomical Observatory (OAO-2) and the European TD-1A, carried out general surveys of the ultraviolet sky. The stars they picked out are fiercely burning suns with temperatures in the range from 10 000 to 100 000 K. As a result, ultraviolet detectors show a very different view of the constellation Orion for example. The prominent red star Betelgeuse is cool on this scale of temperature; with a temperature of 'only' 3600 K it emits very little short wavelength radiation and is invisible in the ultraviolet. The other optically bright star, Rigel, is much hotter: this 11 500 K star is somewhat brighter in the ultraviolet than in the optical. But even Rigel is overshadowed by the brilliance of Orion's belt. Each of these stars is some five times hotter than the Sun, with a temperature around 30 000 K.

Ultraviolet telescopes thus 'see' the hottest stars of all, and because of this they naturally pick out the youngest star groups in the sky. Stars are developing all the time in a

galaxy like our Milky Way. They are formed – invisible to all but infrared detectors – in dark dusty nebulae; but eventually the newly born group of stars blows away the dust and shines undimmed as a new star cluster. The cluster contains stars of all masses, from heavyweights some fifty times heavier than the Sun to diminutive stars of only one-twentieth the Sun's mass. The lightweight stars are dim yellow or red stars, with long lifetimes of several thousand million years or more. During their lives, they can drift away from their birthplaces and spread out around the galaxy. But the heavy stars are profligate with their fuel. They 'burn' much more fiercely, at such high temperatures that they shine most brightly at ultraviolet wavelengths, and they devour their fuel so quickly that they reach the ends of their lives in only a few million years. This is only the blink of an eyelid in the astronomical timescale, so short that the star has no time to travel away from its birthplace before it reaches the end of its natural span and destroys itself in a supernova explosion.

Ultraviolet stars are thus comparatively young stars. They are found only in the youthful clusters of stars which lie close to regions of star birth. The constellation Orion marks the nearest such region in our own Galaxy, and as a result it is bejewelled with ultraviolet stars.

The distribution is more obvious when we turn to galaxies beyond our own. Spiral galaxies are large, disc-shaped conurbations consisting of thousands of millions of stars. New stars are born in the curved 'spiral arms', where gravity clumps the interstellar gas into nebulae. The lightweight stars diffuse away from the arms, to fill the regions in between. Optical telescopes pick up the yellow or reddish light from these stars, and show the whole disc of the galaxy as a gently glowing spiral, shining most brightly at the centre where the old, red stars are concentrated.

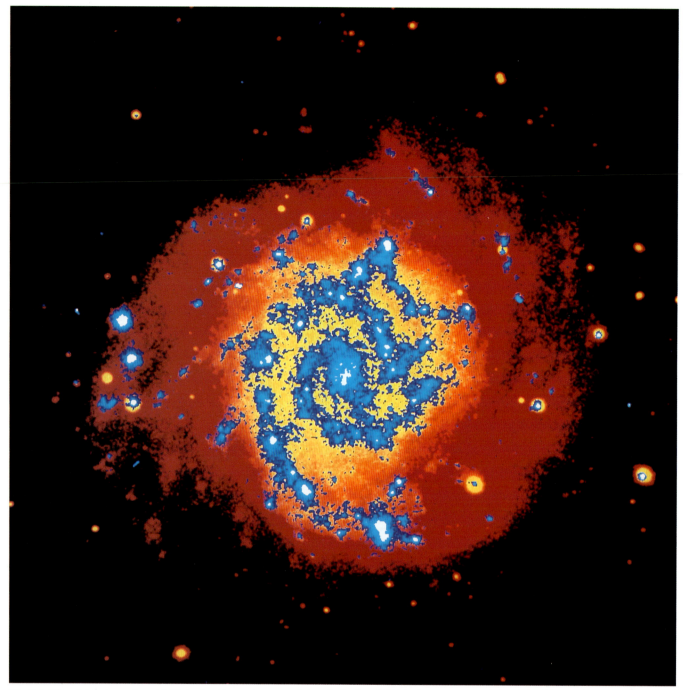

Fig. 9.3 *Ultraviolet observations highlight the thin curving arms of the spiral galaxy M74, in this composite of ultraviolet (blue) and optical (yellow and red) images. The ultraviolet view was obtained by the Ultraviolet Imaging Telescope of the Astro observatory in December 1990. The radiation comes from very hot young stars, still lying in the spiral arms where they were born. The Astro view shows the arms of M74 are more symmetrical than they appear optically: a large-scale pattern of star-formation is apparently sweeping around M74 and imposing a single spiral shape over the entire galaxy.*

Ultraviolet telescopes, on the other hand, are blind to these old stars. They see only the youngest, hottest stars (Fig. 9.3). Because the galaxy's arms are star birthplaces, ultraviolet detectors emphasize the spiral arms and show little of the galaxy's overall structure and inner regions.

Once astronomers had a survey of the ultraviolet sky, the stage was set to study the sources in detail. The second generation of ultraviolet satellites examined the spectra of stars. It included the American-European International Ultraviolet Explorer, IUE (Fig. 9.4).

Astronomers were keen to use their first ultraviolet space observations to study stars' spectra, rather than to produce images, because the ultraviolet spectra contain clues to some of the commonest elements in the Universe. Carbon and nitrogen, for example, do not produce very informative spectral lines at the wavelengths of visible light, but their strong ultraviolet spectral lines show readily the abundance of these elements. Astronomers using IUE have discovered, for example, that a nova explosion in Cygnus in 1978 produced a lot of nitrogen, while the supernova that created the famous Crab Nebula threw out relatively little carbon. These are important clues both to the formation of new elements, and to the mechanisms behind star explosions.

In addition, the radiation from any distant star has to

Fig. 9.4 *The International Ultraviolet Explorer (IUE) satellite was launched in 1978 and switched off 18 years later. IUE, which was in a geosynchronous orbit, could be operated by European astronomers from a control centre in Spain or by Americans in Maryland. The telescope on IUE was not designed to produce images: instead it channeled radiation to two spectrographs which broke up the radiation by wavelength. IUE pioneered astronomical observations in the ultraviolet, from the composition of gas in our Galaxy to the explosion of Supernova 1987A in the Large Magellanic Cloud.*

pass through the interstellar gas on its way to us. This gas between the stars is invisible, and optical astronomy can tell us little about it. But atoms in the gas absorb ultraviolet radiation of specific wavelengths. Observations from IUE show that the gas is far from being uniformly spread: supernova explosions have blasted out 'holes' where the gas is even more tenuous than it is generally in interstellar space, so that the structure of the invisible gas in our Galaxy resembles a Swiss cheese.

Now we are in the era when we have spacecraft sending back ultraviolet images of stars, nebulae and galaxies. The Hubble Space Telescope (Chapter 3), though usually called an 'optical telescope in space', can also observe in the ultraviolet, down to a wavelength of only 115 nanometres. With corrective optics applied to its defective mirror, the space telescope can give its most detailed images at ultraviolet wavelengths – theoretically as fine as 0.04 arcseconds. Its 2.4-metre mirror is far larger than any other ultraviolet telescope, providing unprecedented views of faint and distant stars and galaxies.

A stunning portfolio of ultraviolet images has come from a telescope first flown aboard the Space Shuttle Columbia in December 1990. On this nine-day flight, Columbia carried a set of four telescopes on a common mounting, collectively called Astro (Fig. 9.5). One was an X-ray telescope, the others a complementary trio devoted to the ultraviolet: to image the celestial object under study, to obtain its spectrum and to measure its polarization.

Astro's Ultraviolet Imaging Telescope has a mirror 38 centimetres across, collecting radiation with a wavelength between 120 and 320 nanometres. In many ways, this is a 'scout' for the Hubble Space Telescope. It can photograph a large region of sky – bigger than the Full Moon – so it is ideal for capturing extended objects like star clusters, nebulae and galaxies. Hubble can then home in on any particularly interesting regions in detail.

These ultraviolet instruments consist of telescopes and detectors that are very similar to optical instruments. Ordinary reflecting telescopes with an aluminium reflecting surface will also focus ultraviolet light. The image can be recorded with an electronic detector, or even on ordinary film – provided the film contains a minimum amount of the gelatin used to hold the sensitive silver salts in place, because gelatin absorbs ultraviolet radiation shorter than 220 nanometres.

One very simple kind of image intensifier for ultraviolet radiation is the *microchannel plate* (Fig. 9.6). It consists of thousands of small, very thin, glass tubes stuck together side by side, to form a block that may be a few centimetres across and only a millimetre thick. The tiny tubes form a myriad of parallel channels, each a millimetre long and only one-fortieth of a millimetre in diameter. For ultraviolet astronomy, the plate is housed within a glass envelope which is evacuated to prevent air molecules interfering with

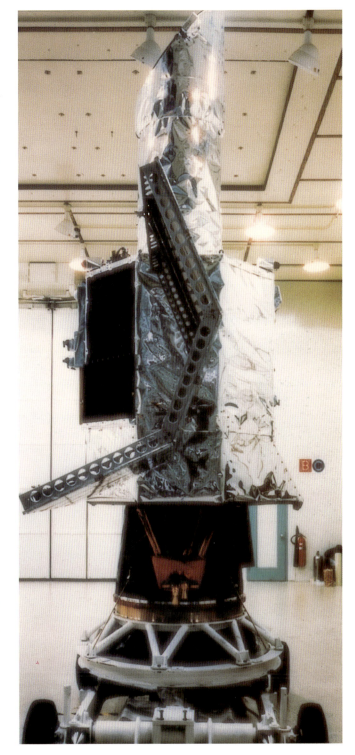

its operation. The front window of the envelope supports a sensitive metal electrode, and the rear window a phosphor screen which produces light when electrons fall on it. A voltage of around a thousand volts is applied across the plate's one millimetre thickness, and a photographic plate is placed at the back to record the final image.

This microchannel plate assembly is fixed at the telescope's focus so that the ultraviolet image falls onto the front electrode. When an individual packet of ultraviolet radiation (a *photon*) hits the electrode, it dislodges an electron into the tube directly beneath. Once inside this microchannel, the electron is accelerated down the tube by

Fig. 9.5 *The Astro ultraviolet observatory peers at the Universe over the side of the space shuttle Columbia during the observatory's maiden flight in December 1990. The four small telescopes check the orientation of the observatory. The large white cylinder (right) is the top of the Ultraviolet Imaging Telescope, which produces views of star clusters and galaxies. Behind are two further ultraviolet telescopes, for studying the spectra and polarization of astronomical sources. A further telescope (out of this shot) observes X-ray spectra. The Astro payload has already flown twice, and further missions are planned.*

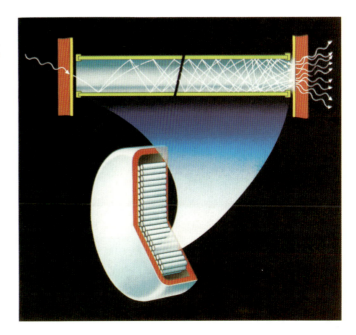

Fig. 9.6 *A microchannel plate consists of thousands of narrow glass tubes side-by-side, in an evacuated glass envelope. Ultraviolet falling on a metal electrode (left) dislodges an electron. A high voltage draws it down the nearest tube, and collisions with the wall multiply the electrons. They hit a phosphor layer (right) to produce a bright spot of light.*

the electric voltage. It does not travel far, however, before it hits the side of the narrow channel. The collision dislodges a few more electrons from the glass wall. These too are accelerated down the tube, and collide with the walls. By the time the first electron reaches the bottom of its microchannel, it is accompanied by an avalanche of around 100 000 electrons knocked out by successive collisions. These electrons plough into the phosphor, and produce a bright spot of light – just as the electrons in a television tube make the phosphor of the screen glow. The spot's position corresponds exactly to the location of the ultraviolet photon in the original image, but it is much brighter. The microchannel plate thus acts as a simple and robust image intensifier, and the image in the phosphor screen is permanently recorded on the photographic emulsion behind.

Photographic plates are a convenient way of recording the intensified image when the whole telescope assembly is destined to return to Earth. They have been parachuted down from balloon and rocket flights, and brought back to Earth by astronauts on the Astro mission. But the detectors on an unmanned satellite must scan the image electronically, so it can be radioed back to Earth like a television picture. Microchannel plates can be adapted for electronic 'reading' by replacing the phosphor and photographic plate with an array of small electrodes (anodes), one at the base of each channel to catch its avalanche of electrons.

Microchannel plates can work right down to the shortest ultraviolet wavelengths, and into the X-ray region. The border between ultraviolet and X-rays – generally set at about 10 nanometres – is in fact rather arbitrary. It arises from the historical accident that radiation produced in the laboratory from gas discharge tubes was called ultraviolet, while the emission produced by firing electrons ('cathode rays') at a target was termed 'X-rays'. Defined in this way, short wavelength ultraviolet in fact overlaps with the longest wavelength X-rays.

There is, however, one natural wavelength marker that occurs in the middle of the ultraviolet region. Atoms of hydrogen – the simplest and most common element – absorb strongly radiation with a wavelength of 91 nanometres or less. Wavelengths from here down to 10 nanometres are called the *extreme ultraviolet* (EUV) – or sometimes the *X-ray-ultraviolet* (XUV) – because they fall on the borderline with X-rays.

The EUV radiation from one particular star has been investigated in remarkable detail. The Sun is sufficiently close that we can make out the finest features of its surface and atmosphere. Since the Sun seems to be a fairly typical star, astronomers hope that a detailed investigation of it can tell us about the structure and lifestyle of stars in general .

Visible light shows us the Sun's 'surface', the *photosphere*, which has a temperature of 5800 K and is a relatively smooth, unruffled ball of light. The Sun's outer

atmosphere, the *corona*, is completely different: it is a complex, turbulent, ever-changing tangle of gas streams. Oddly enough, the corona is very much hotter than the photosphere. With a temperature of a few million degrees, the corona 'shines out' in X-rays. The key to the corona's structure and temperature must lie in the layers between it and the photosphere below.

The layer of gas immediately above the photosphere is the *chromosphere*, visible during eclipses of the Sun as a thin ring of reddish light, emitted by hydrogen atoms. Above the chromosphere lies an extremely thin, and still enigmatic, layer of gas. In this transition region, the temperature rises from about 6000 K at its lower boundary to 1 000 000 K at the top – in a vertical distance of only a thousand kilometres, less than one-thousandth of the Sun's diameter. The gas in the transition region emits radiation at ultraviolet and EUV wavelengths.

The first detailed pictures of the Sun with ultraviolet and EUV detectors came with the unmanned Orbiting Solar Observatory satellites of the 1960s and early 1970s; and the exploration continued with the Solar Maximum Mission satellite. But the most spectacular results came from the American manned space station Skylab, launched in 1973. Three three-man crews inhabited the space station for a total of five and a half months, and amongst their many tasks in orbit was a continual surveillance of the Sun with Skylab's own solar observatory, the Apollo Telescope Mount, or ATM (Fig. 9.7).

The ATM carried eight telescopes which watched the Sun at wavelengths ranging from visible light down through the ultraviolet to X-rays: from 700 nanometres down to just 0.2 nanometres. The three crews took a total of 150 000 pictures of the Sun at different wavelengths.

In the ultraviolet and EUV, our star appears as a blotchy, tempestuous ball of fire (Fig. 9.8). Views at different wavelengths within this region reveal various layers of its chromosphere, transition region and lower corona. Tall prominences of cooler gas stretch lazily upwards, or explode violently into the corona. The corona itself does not exist as a uniform atmosphere, as astronomers had thought before Skylab took its EUV pictures. The hot

Fig. 9.7 *The solar observatory – the Apollo Telescope Mount (ATM) – of the American Skylab space station is the separate structure at the far end, with four solar panels to provide its own power. The various telescopes looked out through the circular face plate.*

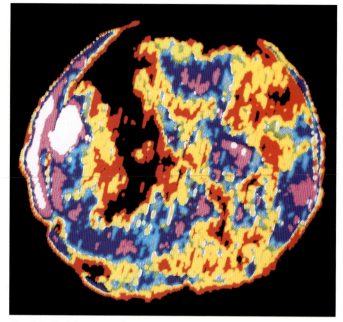

Fig. 9.8 *Skylab's EUV telescopes revealed a blotchy Sun, colour coded here so dim regions are red and successively brighter parts yellow, green, blue, purple, lilac and white. This radiation comes from the transition region at the very base of the Sun's hot atmosphere.*

coronal gas exists only in large individual blobs where it is confined by magnetic fields stretching up from below. In between, there are dark *coronal holes* through which gas streams unhindered into space.

Observing other stars is intrinsically much more difficult than studying the Sun, just because of their immense distances. But in the case of EUV astronomy, there is also a natural fog in space that hides other stars from our gaze. In the early days of ultraviolet astronomy, researchers thought that the hydrogen atoms in space would completely absorb the EUV coming from even the nearest stars – so EUV astronomy seemed like a non-starter.

The absorption occurs because every photon of radiation carries a certain amount of energy, which depends on its wavelength: the shorter the wavelength, the higher the energy. Photons of ultraviolet radiation with a wavelength of 91 nanometres have just enough energy to break apart the proton and electron making up a hydrogen atom, with the photon itself disappearing in the process. Radiation of this wavelength should, therefore, travel only a short distance through space. Ultraviolet photons of slightly shorter wavelength have higher energy, so they too are absorbed by hydrogen in space. These higher energy photons, however, are less likely to hit a hydrogen atom, and as we move towards X-ray wavelengths, we find that photons can once again travel freely through the interstellar gas.

It has turned out, however, that this interstellar 'fog' at the EUV wavelengths is not quite as dense as astronomers had originally feared, and it is sufficiently clumpy that we can see right through it in some directions. Another manned mission was responsible for the discovery of the first EUV sources outside the Solar System. The American Apollo capsule that performed a historic docking with a Soviet Soyuz craft in 1975 carried an EUV 'camera' that picked out a handful of celestial objects.

But progress was slow until the launch of a special survey

instrument in 1990. This British EUV telescope – the Wide Field Camera – was carried piggyback on a large German-American X-ray telescope called Rosat. The EUV telescope had one crucial difference from other ultraviolet instruments. The mirror in a conventional reflector would absorb this short-wavelength radiation, so the Wide Field Camera uses a set of mirrors that reflect radiation at a shallow angle – a design widely used at X-ray wavelengths (see Chapter 11).

Before 1990, astronomers knew of only eight stars (apart from the Sun) that emit EUV radiation. In its six-month survey, the Rosat instrument discovered almost a thousand new EUV sources.

An American satellite, the Extreme Ultraviolet Explorer (EUVE), was launched in 1992 to investigate these sources in more detail. It has mapped the entire sky, sending back images of the larger and brighter EUV sources. These are principally hot shells of gas from stars that have exploded near the Sun, including the Cygnus Loop and the Vela supernova remnant. It is also investigating the spectra of hundreds of EUV sources.

Extreme ultraviolet detectors pick out objects that are even hotter than sources of ultraviolet – with a temperature that is typically 100 000 K. About half the sources discovered in the Rosat survey are ordinary stars surrounded by extremely dense and hot atmospheres, more powerful versions of the Sun's corona.

Most of the other entries in the Rosat catalogue are white dwarfs. A white dwarf is the bare core of a dying star, exposed to space when the star's outer layers waft away at the end of its life. Although a white dwarf does not produce any nuclear energy of its own, it inherits the tremendous heat energy of a star core that was at millions of degrees. Even after cooling for thousands of years, the white dwarf is still hot enough to emit EUV copiously. Where an older white dwarf orbits an ordinary star, gas falling towards the dwarf can heat up enough to produce a flickering output of EUV.

Fig. 9.9 *The Extreme Ultraviolet Explorer (EUVE) satellite, launched in June 1992, fills one of the last gaps in our wavelength coverage of the Universe, between 10 and 91 nanometres. Hydrogen in space is opaque at these wavelengths, but the interstellar gas is sufficiently patchy that EUVE can observe many objects in our Galaxy and even a few distant galaxies. The satellite spins around its long axis, so its trio of survey telescopes (pointing downwards here) scan continuously around the sky. A fourth telescope (right) points along the spin axis and makes longer observations, including spectral analysis.*

But the greatest surprise has been the discovery of EUV coming from the centres of some 'active galaxies'. These galaxies have a tiny brilliant centre, producing enormous amounts of radio waves, X-rays and gamma rays. Most astronomers believe this radiation comes from a disc of gas on the brink of a massive black hole. The disc should certainly shine brightly at EUV wavelengths, but researchers had thought this radiation would be absorbed by gas clouds further out in the active galaxy – and any radiation that did escape would be mopped up by hydrogen in our Galaxy on its last lap towards the Earth.

Now, it turns out that there are some active galaxies where we are doubly lucky. The gas clouds in the galaxy are arranged so that we can see between them, through a 'window' into the galaxy's very core. And we see the galaxy through a comparatively empty 'window' in the gas of our own Galaxy, that lets this radiation through to us.

In the 1990s, ultraviolet astronomy – including the EUV wavelengths – came of age. After years of being able to study only the stars of our own portion of the Milky Way, ultraviolet astronomers are now coming face to face with the extragalactic: other nearby galaxies and even exploding galaxies deep in the Cosmos.

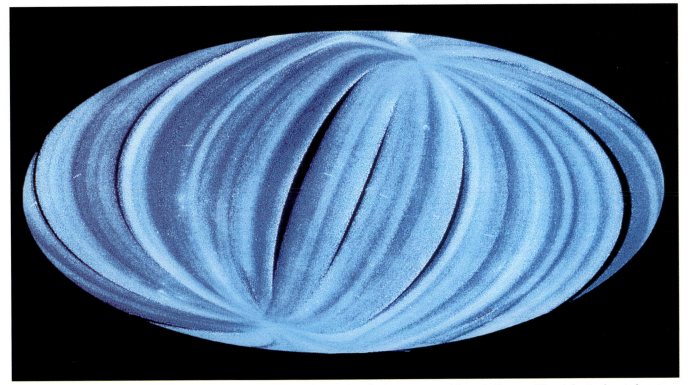

Fig. 9.10 *This pioneering view of the entire sky at extreme ultraviolet wavelengths was constructed from the Extreme Ultraviolet Explorer's first six months of observation. The satellite scanned the sky in strips running from near the North Pole (top) to the South Pole (bottom). The brighter strips here were more intensely scrutinized by EUVE, and this scanning pattern dominates the image. But the brighter astronomical sources can be made out as well. The moving Moon shows up as the dashed line running across the image. Circular features are supernova remnants: Vela at lower left and the Cygnus Loop at upper right. The hot stars of Orion, including the distinctive shape of the Belt and Orion Nebula, are seen at centre left.*

10 normal galaxies

BEYOND THE MILKY Way lie an estimated 100 thousand million other galaxies. Although many had been catalogued as 'nebulae' in the past, it is only during the twentieth century that their true nature was realised. Much of the credit must go to Edwin Hubble who, in the 1920s and 1930s, systematically photographed hundreds of galaxies and classified them into different types according to their appearance.

Hubble's classification has stood the test of time. Although the original scheme has been extended, refined and subdivided, astronomers still classify galaxies into three basic types – spiral, irregular and elliptical – because a galaxy's appearance really does reflect its internal conditions. (Some spiral and elliptical galaxies also have small, very active cores; they are the 'active galaxies' covered in Chapter 12.)

Spiral galaxies resemble our Milky Way in both size and form. All have a central bulge made of older stars, surrounded by a disc of young stars, and dust and gas in which the spiral arms are embedded. Some spirals have bar-shaped rather than circular bulges ('barred spirals'), and there are individual variations in the smoothness of the arms and the degree to which they are coiled around the bulge. Each spiral galaxy has its own individuality, but all are obviously part of the same family.

Irregular galaxies, like the Small Magellanic Cloud, do not share this coherence. Although they resemble the spirals in having a similar make-up of gas, some dust, and young stars, they are smaller and have no particular shape.

Elliptical galaxies are quite unlike the other two types. They have used up virtually all their dust and gas in star formation. As a result, they are smooth, featureless balls of old, red stars almost devoid of interstellar matter. What they lack in individuality some compensate for in size, however, and some elliptical galaxies are among the biggest and most massive in the Universe. But they can also range down to the other extreme. Some dwarf ellipticals contain less than a million stars and are scarcely visible unless they lie close to us. These are probably the commonest kind of galaxy in the Universe.

Galaxies can exist singly, but it is more common to find them in pairs, groups or clusters, held together by their mutual gravitational attraction. All the galaxies described in this chapter belong to small groups of a few dozen members, most of them dwarfs. At the other extreme are the giant clusters of galaxies, in which hundreds or thousands of galaxies – mainly ellipticals – cover volumes of space up to 50 million light years in extent.

Fig. 10.1 is an optical photograph of a typical spiral galaxy, NGC 253. Its designation means that it is the 253rd object in the New General Catalogue of Nebulae and Clusters of Stars, published in 1888 by Johan Dreyer. Other bright, nearby galaxies are prefixed by an 'M', which indicates that they were included in the catalogue published by comet-hunter Charles Messier in 1784 listing objects liable to be confused with comets.

NGC 253 belongs to the Sculptor group of galaxies which, at just under 10 million light years away, is the nearest small cluster to our Local Group and very similar to it in size. With a diameter of 40 000 light years, NGC 253 is the group's largest member. Although less than half the size of the Milky Way, NGC 253 is, nevertheless, four times brighter and, with a mass of 200 thousand million Suns, just as massive. It must look very similar to our Galaxy, but because we see it from only 17° above its rim, we miss a lot of its details. However, Fig.10.1 clearly shows the smooth yellow glow from the central bulge, contrasting strongly with the dusty blue spiral arms. Bands of dust are particularly obvious on the North (top) side. This is the edge of the galaxy closest to us, and the disc's thick dust lanes can be seen silhouetted against more distant star-clouds.

Optical photographs tell only part of the story, however. In the infrared, NGC 253 is the third brightest galaxy in the sky. This is because – like the galaxy M82 – it is unusually rich in dust, which absorbs light from stars and re-radiates it at longer wavelengths. At radio wavelengths, it has a central source which is a hundred times more powerful than its counterpart in our Galaxy, and a thousand times more powerful than that in the Andromeda Galaxy. X-rays tell of a recent burst of star formation near the centre, an event given additional support by radio and infrared observations. Yet little of this activity is visible at optical wavelengths.

The same is true of all galaxies. Appearances are not always as they seem. Today, multi-wavelength observations are 'peeling the layers off galaxies' as never before, to reveal that even the most ordinary have some extraordinary features.

10.1 NGC 253. Optical, true colour, 3.9 m Anglo-Australian Telescope

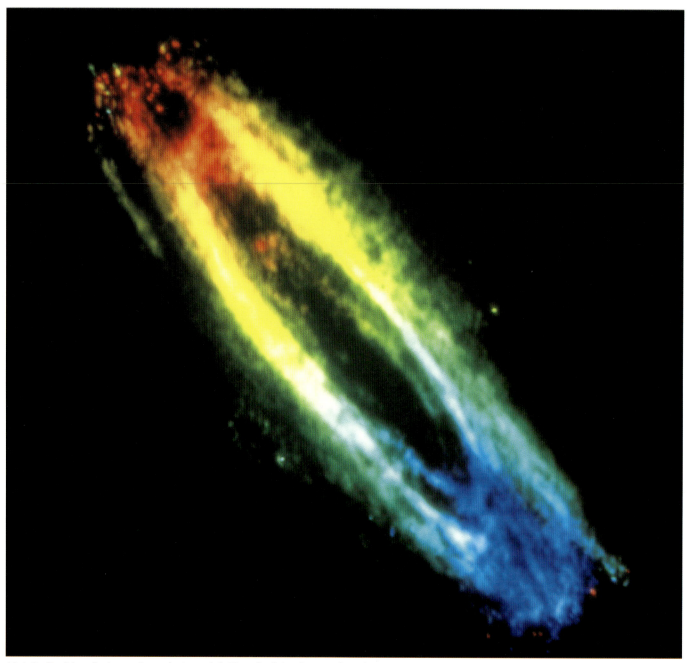

10.4 *Radio, 21 cm hydrogen line, velocity coded, Westerbork Synthesis Radio Telescope*

Like the Milky Way, M31 is a galaxy whose spiral arms are rich in cold hydrogen gas. The gas emits its characteristic signal at a wavelength of 21.1 centimetres, enabling astronomers to map its distribution and its movement by radio telescopes. Observations at this wavelength reveal a great deal about the dynamics of the galaxy.

Mapping the Andromeda Galaxy at 21 centimetres is not entirely straightforward. M31 has an intrinsic velocity of its own within the Local Group, which is superimposed on all the motions of gas within it. In fact, M31 is one of the rare galaxies which are approaching us, rather than rushing away; it lies so nearby that its motion (with respect to the Milky Way) is not affected by the expansion of the Universe. The net velocity of approach of M31 is 310 kilometres per second – which, despite its apparently high value, means that the two galaxies will move closer together by only 0.0004 per cent in the next million years!

Fig. 10.4 is a radio image of the hydrogen in M31, colour coded for velocity, and corrected for the intrinsic motion of the galaxy itself. The brightness of the emission reveals the amount of gas present, while the colour shows its speed towards or away from us: green and blue regions are approaching while yellow and red regions are receding.

Although Fig. 10.4 shows the same extent of the galaxy as the optical photograph (Fig. 10.2), there is a striking reversal of M31's appearance. The optically dominant central bulge is dark at these wavelengths, while the formerly inconspicuous rim of the galaxy shines brilliantly at 21 centimetres. The glowing regions here reveal the galaxy's cold hydrogen reserve. Towards the central bulge, the zone dominated by old red and yellow stars, all the hydrogen has already been used up, or is in the form of hydrogen molecules (H_2). But in the disc there is still a considerable number of hydrogen atoms, which show up particularly where the gas has been compressed by the passage of a spiral arm. For instance, strong concentrations here correspond to the dust lanes on the right of and above the central

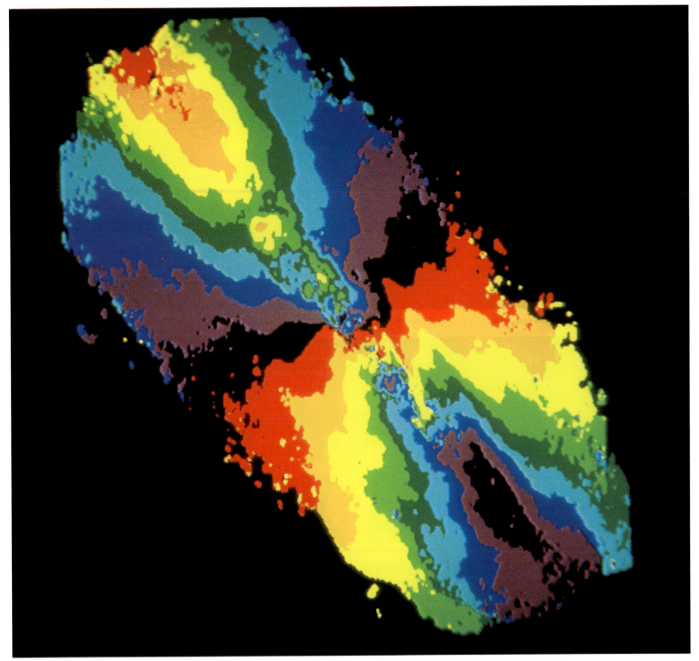

10.5 *Radio, 21 cm hydrogen line, velocity coded (irrespective of intensity), Westerbork Synthesis Radio Telescope*

bulge; and gaping 'holes' occur in the outermost regions where there are spaces between the spiral arms.

The velocity colour coding shows that M31 is in a state of systematic motion. The lower portions are approaching, while the upper parts are receding, revealing that the whole galaxy is rotating in space with its spiral arms trailing behind it. This is shown in more detail in **Fig. 10.5**, which codes the velocity of hydrogen gas right across the galaxy, irrespective of its amount (and hence its intensity at 21 centimetres). The black contour shows gas across the minor axis, which is effectively at rest. The other colours are at velocity intervals of approximately 25 kilometres per second: towards the lower right, red, orange, yellow, green,

blue, and purple regions show increasing velocities of approach; while to the top left the opposite sequence of colours indicates increasing speeds of recession.

Two intriguing features emerge from this 'scarab beetle' plot. First, the contours are not smooth parabolas pointing towards the galaxy's centre – which they would be, if the Andromeda Galaxy were rotating smoothly. There are 'bumps' of anomalous velocity throughout, such as the yellow and purple patches just above and below the centre. These correspond to the positions of the spiral arms, and they may be a result of the 'density wave' (which creates the arms) disturbing the speed of the gas as it passes through.

Second, the way the velocity changes

with distance from the centre (the rotation curve) enables us to find the mass of M31, because the rotation speed at a given radius is dictated by the amount of matter lying within. The rotation speed should increase with radius near the centre, but after a maximum it should then fall off gradually with distance. Fig. 10.5 shows that the outermost parts of M31 have a definite tendency to maintain a high rotation rate, implying that there must be some unaccounted mass whose gravitational field holds the high-speed outer arms in check. Astronomers suspect that this mass lies hidden in a huge dark halo which surrounds the Andromeda Galaxy, and contains ten times the mass of the visible galaxy.

10.15 *Optical, enhanced true colour, 1.2 m Palomar Schmidt Telescope*

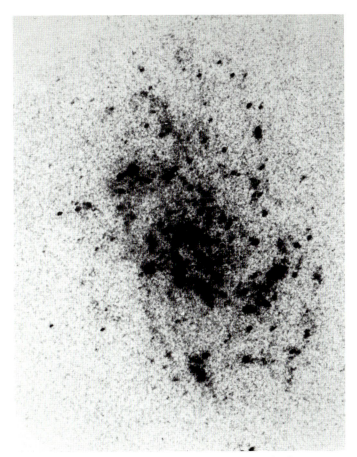

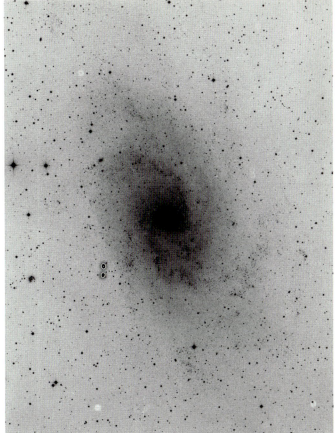

10.16 *Ultraviolet, 125–175 nm, negative print, 0.3 m rocket-borne telescope*

10.17 *Optical, 710–880 nm, negative print, 1.2 m Palomar Schmidt Telescope*

M33

M33 LIES CLOSE to the Andromeda Galaxy in the sky, in the neighbouring constellation of Triangulum, and it is another major member of our Local Group of galaxies. At 2.5 million light years, slightly further away than Andromeda, it is sometimes claimed to be the most remote object visible to the unaided eye. However, M33 is much fainter (with an apparent magnitude of 6.5) and not visible unless sky conditions are exceptionally good.

The optical view of M33 in **Fig. 10.15** is enlarged to twice the scale of the similar view of the Andromeda Galaxy in Fig. 10.2. Even allowing for M33's greater distance, it is obvious that the Triangulum galaxy is much smaller than its giant neighbour. With a diameter of 40 000 light years and a mass of 15 000 million suns, it is much more similar in size to the Large Magellanic Cloud (Fig. 8.19), and like the Large Magellanic Cloud, M33 is vigorously forming new stars, which show up in Fig. 10.15 as ragged, blue spiral arms.

In this computer-enhanced optical picture, photographs taken through blue, green and red filters have been combined with their respective colours printed as strongly as possible, to emphasise the slight colour differences in M33. Star formation

can be seen in several giant clouds of glowing hydrogen – vivid pink in Fig. 10.15 due to emission from hydrogen atoms at 656 nanometres – especially those strung out along the spiral arm which unwinds from just right of centre and extends upwards to the left. Most prominent of all is NGC 604, at the end of the arm. In size, mass and appearance it is very similar to the Large Magellanic Cloud's enormous Tarantula Nebula (Fig. 8.24).

M33 is a more 'open' spiral than M31 or the Milky Way, with wide, far-flung arms and a proportionally smaller centre bulge. Mixed in with the young stars is a good deal of dark interstellar dust, which shows up as reddish-brown – its actual colour – against the underlying yellow glow from the older stars in the galaxy. The latter's smooth distribution contrasts strongly with the patchy spread-out pattern made by the young stars.

Fig. 10.16 shows an ultraviolet view of M33. In contrast to similar pictures of M31 (Figs. 10.9 and 10.10), where the emission is restricted to an outer 'ring', there is ultraviolet emission from all over the galaxy. In fact, the ultraviolet view is little different from the view at optical wavelengths. Both are dominated by the radiation from hot young O and B stars, which have been recently formed

throughout the whole galaxy.

Fig. 10.16 also reveals the true extent of M33, highlighting clumps of O and B stars at great distances from the centre. The emission is strongest in regions where stars have already formed (below centre). It is, however, weak in areas where stars are still forming, such as in the nebula-studded spiral arm above centre, because the dust absorbs ultraviolet radiation from the embedded stars. NGC 604 is hence not very prominent at ultraviolet wavelengths.

In complete contrast, **Fig. 10.17** was obtained at the longest optical wavelengths, and this 'far red' view shows a less extensive, much smoother distribution of stars. (The two spots to the left of M33 are photographic defects.) The main body of the galaxy at this wavelength is half the size of the ultraviolet image. But the older, red and yellow stars which emit this radiation are the types which make up by far the greater proportion of the galaxy's mass; the young, blue, spiral arm stars amount to comparatively little. The old stars, and hence the galaxy's mass, are concentrated into the prominent central bulge. Silhouetted against it are the patchy dust lanes, while the surrounding disc shows only smooth, spiral arms in contrast to the narrow, patchy zones of recent star formation so dominant in Fig. 10.15.

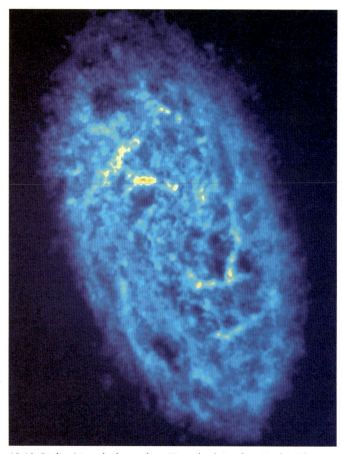

10.18 *Radio, 21 cm hydrogen line, Westerbork Synthesis Radio Telescope*

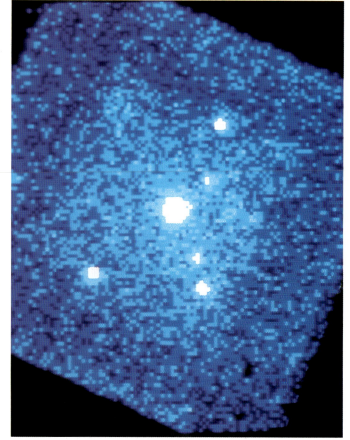

10.19 *X-ray, 0.3–2.5 nm, Imaging Proportional Counter, Einstein Observatory*

Vast reserves of interstellar gas, waiting to fuel M33's rash of star-formation, appear in **Fig. 10.18**. It shows the distribution of cold hydrogen atoms, colour coded to indicate intensity and hence the amount of interstellar gas. The coding runs from dark blue for weak emission, through light blue and yellow to red for the most intense regions.

While the Andromeda Galaxy concentrates its cold hydrogen into a well-defined ring (Fig. 10.4), the cold interstellar gas in M33 occurs throughout the galaxy, right up to the centre. The brightest (red) region consists of gas around the giant nebula NGC 604. To the upper left of this nebula is a line of dense hydrogen clouds (yellow), without any counterpart at optical or ultraviolet wavelengths: it is evidently earmarked for starbirth in the not-too-distant future.

Belying the ragged appearance of M33 at optical wavelengths, Fig. 10.18 clearly shows the interstellar gas marshalled into spiral arms. Each arm is around 1000 light years wide, and some extend for 12 000 light years before their structure becomes jumbled. But there is no simple pattern to this spiral. Unlike the elegant two-armed whirl of M51 (Fig. 10.21) or M81 (Fig. 10.36), M33 consists of dozens of separate 'arm fragments'.

This structure is almost certainly a result

of the energetic star formation in M33. Massive stars are born in the spiral arms, and explode within a short period – cosmically speaking – as supernovae, which compress the surrounding gas. Because the galaxy is rotating, these zones of compression are drawn out into long spiral shapes. This process, 'stochastic self-propagating star formation', demonstrates the complex interactions within a single galaxy. The cold hydrogen gas is the raw material of stars; yet, as the gas is converted into stars, it affects the remaining reservoirs of star-material.

The power of the young stars in M33 is stunningly portrayed in **Fig. 10.20**. This montage of negative photographs was taken with the Soviet 6-metre reflector, until 1990 the world's largest telescope, through a filter that isolated the light from hot hydrogen. It covers the central regions of M33 (for reference, NGC 604 is at the upper left), revealing intimate detail in its hundreds of nebulae.

The galaxy's spiral structure is almost lost here in a chaos of loops and filaments. As well as lighting up the nearby gas as nebulae, the hot energetic stars of M33 are inflating huge glowing bubbles in the interstellar medium. Some are well defined, while others are broken or merge with their neighbours. The largest bubbles are 900 light years across.

This most detailed view of glowing gas in another galaxy has another lesson for galaxy researchers. There is a continuous range in size, shape and brightness from the brilliant NGC 604 and the largest bubbles to the faintest scrap of nebulosity. A galaxy does not consist of isolated nebulae separated by vast reaches of featureless cold gas: the interstellar medium is interconnected and dynamic.

A couple of the bright nebulae associated with starbirth show up at X-ray wavelengths (**Fig. 10.19**). Most of the other X-ray sources here are indicators of stardeath, their emission powered by gas falling towards neutron stars or black holes.

But far and away the most brilliant X-ray source in M33 lies at its very centre. It is a remarkably strong source for such a small galaxy, at least a thousand times more powerful than the X-ray source at the centre of the Milky Way. The X-rays from this central powerhouse can also vary over a period of just a few months. This is a very rapid flickering on the galactic scale, and indicates the X-ray source is less than a light year across. It bears all the hallmarks of the tiny energetic 'engines' found in the centres of active galaxies (Chapter 12), but the light from M33's innermost core shows no signs of activity, nor is there a detectable radio source – usually the most obvious sign of an active galactic nucleus.

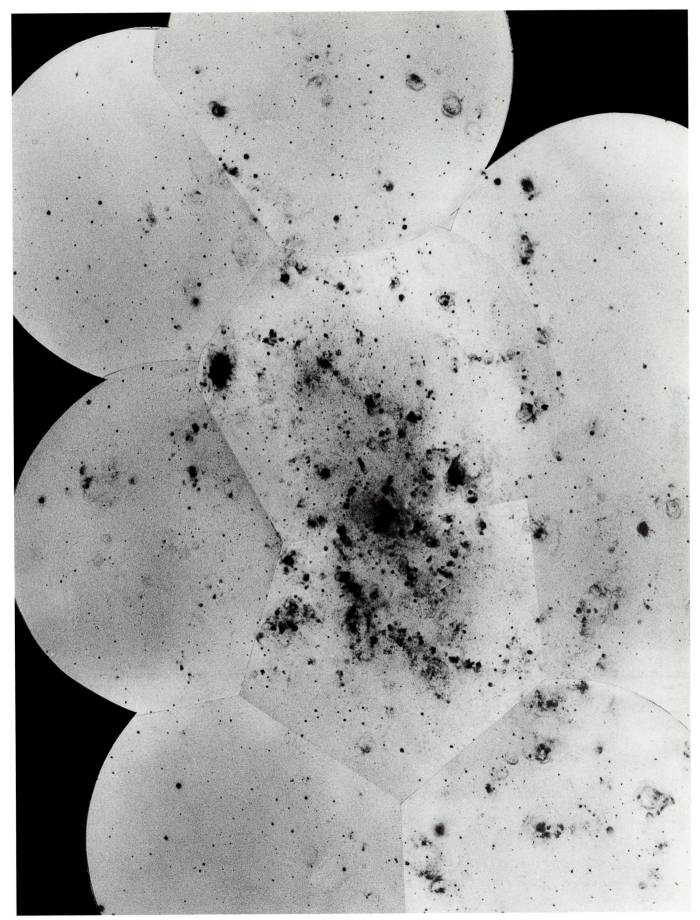

10.20 *Optical, 656 nm hydrogen line, mosaic of negative prints, 6 m reflector, Zelenchukskaya Astrophysical Observatory*

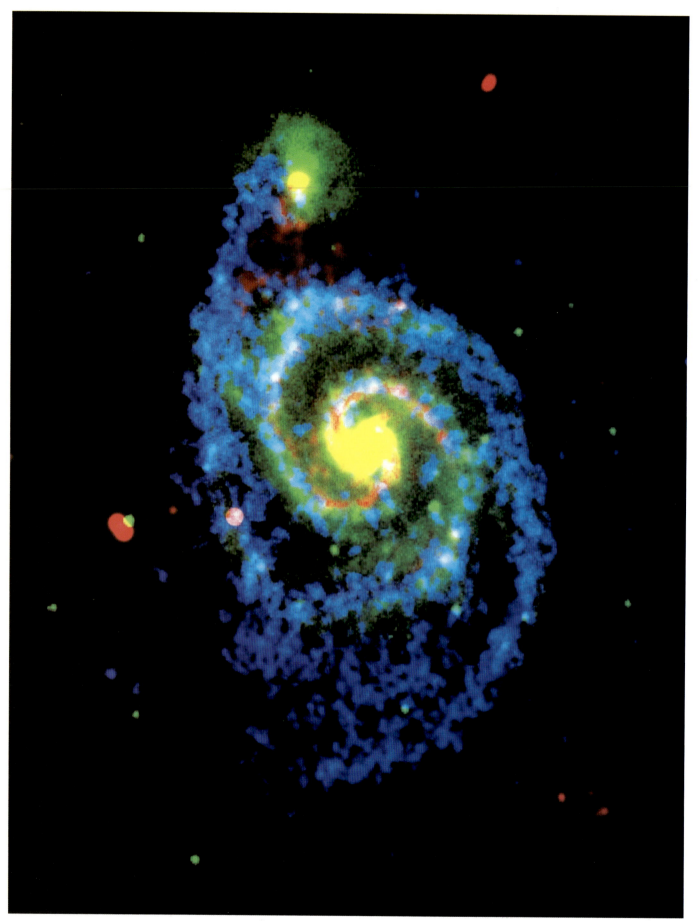

10.21 *Combined wavelengths: optical (green), 1.2 m Palomar Schmidt Telescope; radio, 21 cm hydrogen line (blue) and 21 cm continuum (red), VLA*

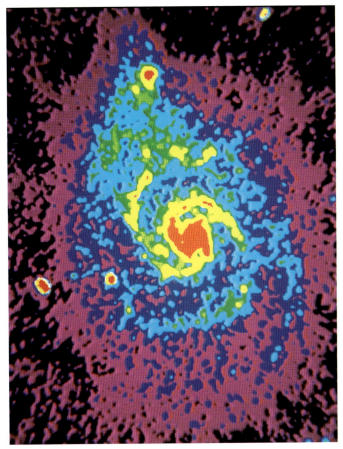

10.22 *Optical, true colour, 1 m reflector, US Naval Observatory*

10.23 *Radio, 21 cm continuum, Very Large Array*

M51

'SPIRAL CONVOLUTIONS; . . . WITH successive increase of optical power, the structure has become more complicated . . . ' So wrote the third Earl of Rosse about this galaxy in 1845, after he observed it with his monster telescope, located at Birr in Ireland. He was the first astronomer to discover spiral structure in galaxies, and he concluded that M51 was 'the most conspicuous of the spiral class.'

To this day, M51 has continued to hold centre stage in the quest to understand the shape of spiral galaxies. It is the pre-eminent example of a 'grand design' spiral. While the spiral arms of the Andromeda Galaxy more resemble a ring (Fig. 10.4), and those of M33 a patchwork of fragments (Fig. 10.18), M51 has a clear pattern of two bold spiral arms.

The shape which gives M51 its nickname 'The Whirlpool' is prominent at all wavelengths. **Fig. 10.21** is a composite of an optical image (green) with two radio observations: the blue image traces the distribution of cold hydrogen gas, while the red image shows emission from fast electrons whirling in magnetic fields. All reveal the same pair of spiral arms, with the interstellar gas (blue) extending much further out than the distribution of stars (green).

The overall diameter of M51, as revealed

by the cold hydrogen in Fig. 10.21, is 65 000 light years, and the galaxy's mass amounts to 50 000 million suns. The Whirlpool is both smaller and less massive than our Galaxy, yet it is four times brighter overall.

The optical photograph (**Fig. 10.22**) reveals clearly that the object to the North (top) of M51 is a separate galaxy, catalogued as NGC 5195. Even this small companion is three quarters as luminous as our Milky Way. As a result, the M51 pair are easily visible in a small telescope from Earth, even though they lie 20 million light years away.

The radio emission is seen more distinctly in **Fig. 10.23**, a colour-coded image where the brightest regions are red, with intensity decreasing through yellow and green to blue and purple. There is a bright radio source at the Whirlpool's centre – one of the strongest in any normal galaxy – and a much less powerful one in NGC 5195.

The most prominent radio spiral arms (yellow and red in Fig. 10.23) are much narrower than the broad arms of stars seen at optical wavelengths. And the composite image in Fig. 10.21 reveals that the prominent radio arms (red) lie on the *inside* of the optical arms (green).

These observations all help to confirm the 'density wave' theory for the arms in M51 and other grand design spiral

galaxies. According to this theory, the mass of stars concentrated in the two arms creates a spiral-shaped pattern of gravitational force. This spiral density wave keeps its identity for hundreds of millions of years, even though individual stars and gas clouds move in and out of the arms during this period. It's like a traffic jam on a motorway that can persist for hours after the original reason for the hold-up has been cleared, with individual cars entering and leaving the jam all the time.

The density wave corrals all the matter in the galaxy – stars and gas – into the same double spiral shape. This contrasts with flocculent galaxies like M33, where the stars, cold hydrogen and nebulae are scattered much more indiscriminately (Figs. 10.15, 10.18, 10.20).

According to theory, the density wave pattern should be rotating comparatively slowly. Stars and gas should catch up with the spiral arm at the rear, its inside edge. Like cars in the traffic jam, the interstellar gas now has to slow down suddenly and is strongly compressed. The magnetic field in the gas is squeezed, too, so its output of radio waves should increase markedly along a narrow band inside each arm – just as observed. The narrow band of red in Fig. 10.21 shows, in effect, the brake lights of the interstellar gas as it encounters the density wave jam on the galactic highway.

The bold 'grand design' of the Whirlpool Galaxy and its unusual brightness can both be traced to the influence of its companion, NGC 5195. And, surprisingly, the effects of this small neighbour can be seen even at the very core of M51.

The companion galaxy shows up prominently in an infrared image (**Fig. 10.24**), colour coded so that bright regions are white and dimmer parts successively yellow and red. Individual small dots are foreground stars in our Galaxy, while a multitude of faint cool stars delineate the outlines of M51 and NGC 5195.

In an optical view (Fig. 10.22), NGC 5195 is half hidden by dense swathes of dust, and seems to have a fairly irregular shape. But infrared radiation can cut through dust, and Fig. 10.24 reveals its true proportions. This companion turns out to be surprisingly symmetrical: it is a small barred spiral galaxy.

Computer simulations of these two galaxies suggest NGC 5195 is orbiting M51 with a period of 500 million years. The smaller galaxy skimmed the top of M51 about 70 million years ago, and is now receding behind the larger galaxy. According to these calculations, NGC 5195 actually lies well behind M51's upper arm that seems to link the two.

This close brush with NGC 5195 was almost certainly responsible for the shape of the Whirlpool Galaxy. According to the density wave theory, the grand design in a galaxy like M51 is self-sustaining – once it has been started off. But a powerful stirring force is required to get the pattern going, and the gravity of NGC 5195 would fit the bill precisely.

The strong gravitational field of the arms in M51 sweeps up and crushes interstellar matter with high efficiency. The gas and dust condense into dense clouds that spawn clusters of hot massive stars, shining like pearls on a necklace when viewed at ultraviolet wavelengths (**Fig. 10.26**). This image is colour coded for intensity, with the dimmest regions red and brighter parts of the galaxy white, blue, green and yellow.

In the ultraviolet, the companion galaxy NGC 5195 is practically invisible. It has a dearth of hot O and B stars, and dense dust clouds that blanket ultraviolet radiation even more effectively than light.

But the star formation it has provoked in M51 is rampant – especially in the arm that points towards the companion, and in a corresponding position on the opposite side. Star formation has proceeded vigorously in these outer parts of the galaxy, and has yet to spread to the inner parts where the radio arms (Fig. 10.23) are brightest. Gas in the galaxy's very heart, though, has also been triggered into a burst of star formation, giving M51 an ultraviolet-bright centre.

A very high-resolution infrared image (**Fig. 10.25**) shows the distribution of stars

10.24 *Infrared, 2.2 μm, 3.8 m UK Infrared Telescope*

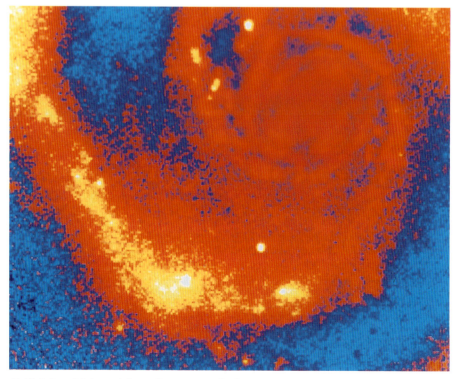

10.25 *Infrared, 2.2 μm, 2.3 m reflector, Steward Observatory, Kitt Peak*

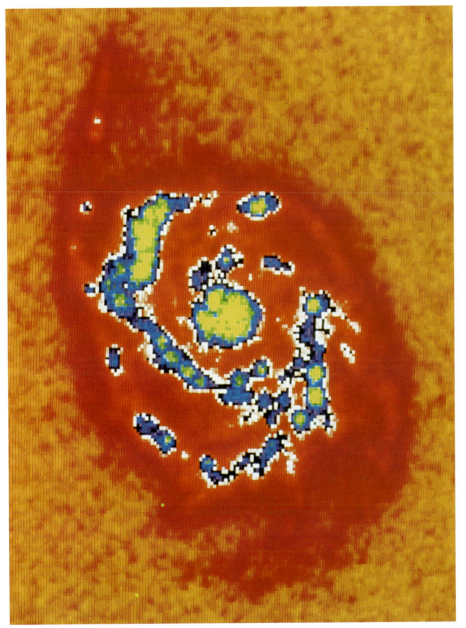

10.26 *Ultraviolet, 200 nm, 0.4 m balloon-borne telescope*

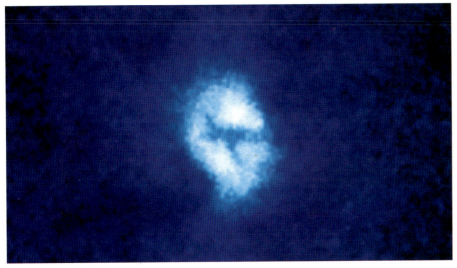

10.27 *Optical, Wide Field/Planetary Camera, Hubble Space Telescope*

at the centre of the Whirlpool in unprecedented detail: optical views are confused by a scattering of bright star clusters and obscured by patches of dust.

Fig. 10.25 is colour coded for infrared brightness, with the most intense regions white, fainter parts yellow and red, and the dimmest regions blue. It covers 15 000 light years of M51, to twice the scale of the previous infrared image: to compare the two, the bright arm to the lower left of the galaxy's centre in the wide-angle view (Fig. 10.24) is also the most brilliant region in Fig. 10.25. The galaxy's centre is to the upper right, and appears in subdued shades of red and blue because the overall brightness of the galaxy's smooth central bulge has been subtracted to reveal small-scale features in the distribution of stars.

The stars in the central bulge, it turns out, are arranged in a double spiral that curves right in to a short central bar, just 2000 light years long. This pattern is a continuation of the twin spiral seen in the outer part of the galaxy. From the galaxy's outer edge to the centre, each arm can be traced for three continuous revolutions – twice as many as observed in any other spiral galaxy. It gives further validity to the Whirlpool's traditional name!

At face value, this discovery is a clear vindication of the density wave theory, which predicts that spiral waves should propagate inwards from the outer arms towards the galaxy's centre. But it also raises problems. At a certain radius from the galaxy's centre, the stars are orbiting at same speed as the spiral pattern. In theory, a long-lasting density wave should not be able to propagate inside this Lindblad resonance (named for the Swedish astronomer Bertil Lindblad).

Perhaps the spiral pattern we see in M51 is only transitory, and will smear itself out into a ring. Or the stars may be moulded by the pull of a spiral configuration of gas. This is the reverse of the usual situation, but the centre of M51 is unusually rich in interstellar material – some condensing into the brilliant young stars seen in the ultraviolet view (Fig. 10.26).

The penetrating eye of the Hubble Space Telescope has shown the fate of interstellar material that strays right into the centre of M51. The bright patch in **Fig. 10.27** is a cluster of stars, only 100 light years across, at the heart of the Whirlpool. The dark horizontal band is a ring of gas and dust that is probably orbiting a black hole as heavy as a million Suns. Bright patches above and below are jets of energetic particles accelerated by energy from the black hole's gravity. They are also responsible for the strong radio emission (Fig. 10.23) from the centre of M51. Although M51 is undoubtedly a 'normal galaxy', its core more resembles a miniature version of an active galaxy like Centaurus A (Fig. 12.2).

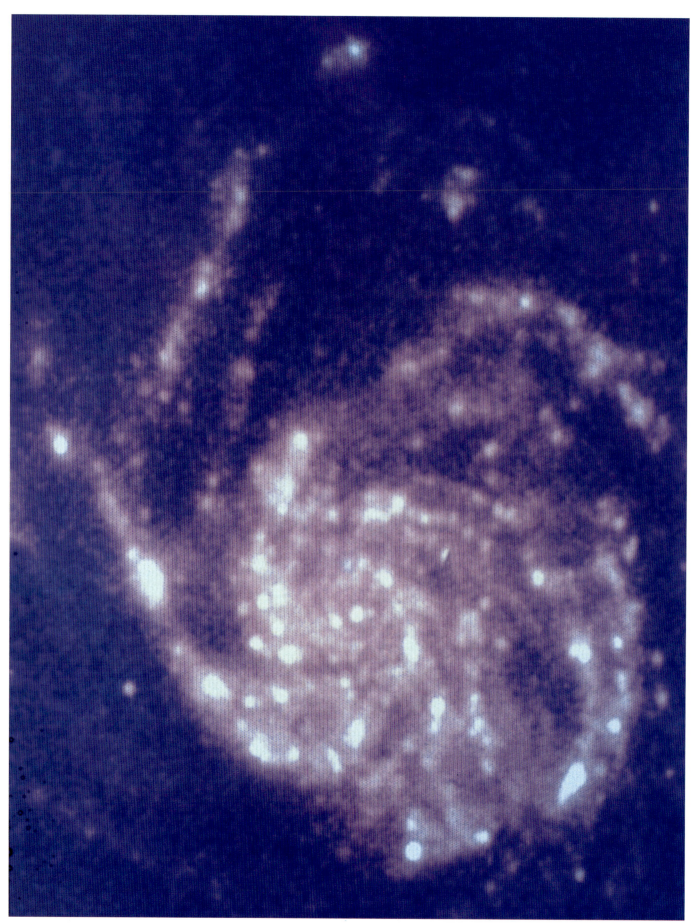

10.28 *Ultraviolet, 200 nm, 0.4 m balloon-borne telescope*

10.29 *Optical, 3.8 m Mayall Telescope*

M101

SPANNING NEARLY HALF a degree of sky in the constellation of Ursa Major, M101 (**Fig. 10.29**) is one of the largest and brightest spiral galaxies known. At its distance of 20 million light years, this angular size corresponds to a diameter of 200 000 light years – twice that of our Milky Way.

M101, the 'Pinwheel Galaxy', has a mass of three thousand million Suns and belongs to a loose group of galaxies which also includes M51 (Fig. 10.21). Its extremely narrow, far-flung spiral arms are dominated by huge clouds of glowing hydrogen, some

of which are 3000 light years across. A number of them in the outer arms in Fig. 10.29 are so bright that they have been allocated their own NGC numbers.

Because M101 is so rich in young hot stars, it is very bright at ultraviolet wavelengths (**Fig. 10.28**). The brightest regions are coded white in this map, with the intensity decreasing through pink to blue. There is a prominent band of young stars to the left of centre, and other concentrations along the inner arms. This ultraviolet image also reveals the true extent of the galaxy, well beyond the limits visible on optical photographs. The bright

nebula at the left of the optical view (Fig. 10.29), for example, is at an 'elbow' in the ultraviolet image: two parallel long straight portions of spiral arm stretch up from it, almost invisible to ordinary telescopes.

The emission from these faint 'ultraviolet arms' does not, however, come directly from young stars, which would be easily visible at optical wavelengths. Instead, it is radiation from other parts of the galaxy which has been scattered by dust. The dust, with cold hydrogen gas (Fig. 10.32), forms invisible extensive reserves in the outer arms.

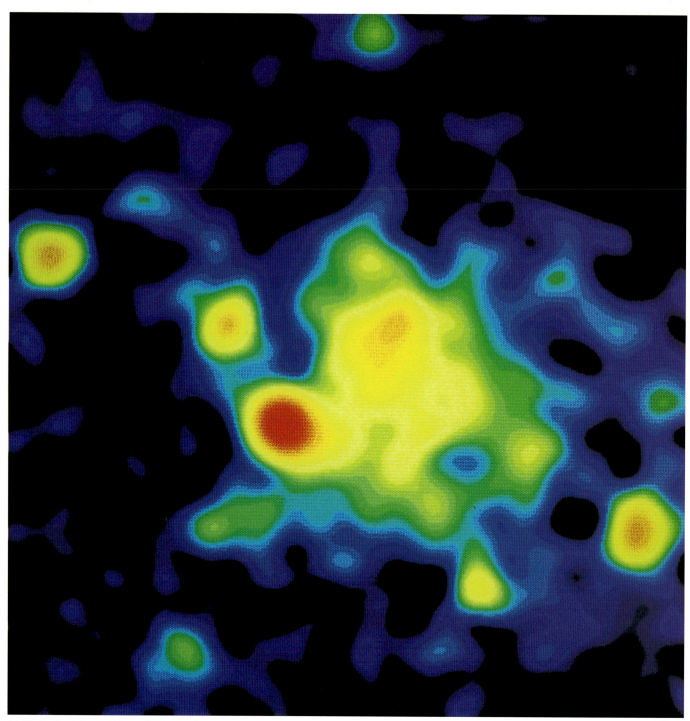

10.30 *Radio, 2.8 cm, 100 m Effelsberg Telescope*

At short radio wavelengths (**Fig. 10.30**), the appearance of M101 is dominated by strong emission from its gigantic hot nebulae. This colour-coded image shows the highest intensities as red, decreasing through orange, yellow, green and finally blue. The spiral arms can just be picked out, but both they and M101's central source are weak at a wavelength as short as 2.8 centimetres. They emit synchrotron radiation as electrons whirl around in magnetic fields, and this kind of radiation is naturally strongest at long radio wavelengths and drops off steadily as you observe at shorter wavelengths.

But the nebulae in M101 stand out brilliantly. Hot gas emits radio waves with equal intensity over all wavelengths, so nebulae become more prominent at short wavelengths where the synchrotron emission from other sources becomes weaker. The brightest radio sources here match up easily with the most powerful nebulae seen in the optical photograph (Fig. 10.29). The nebula to the lower right in the radio view (NGC 5447) is also in the lower right corner of Fig. 10.29. The radio-emitting nebula at far left is out of the frame of the optical photograph, but the next in (NGC 5462, an orange peak within a blue spiral arm) corresponds to the nebula at the 'elbow' in the optical and ultraviolet (Fig. 10.28) images. The radio image reveals that the next nebula along (NGC 5461, red peak) is the most powerful – although dust dims its true importance at optical and ultraviolet wavelengths.

With an average diameter of 3000 light years, these nebulae are ten times larger than their counterparts in the Milky Way, and give out twenty times more energy at

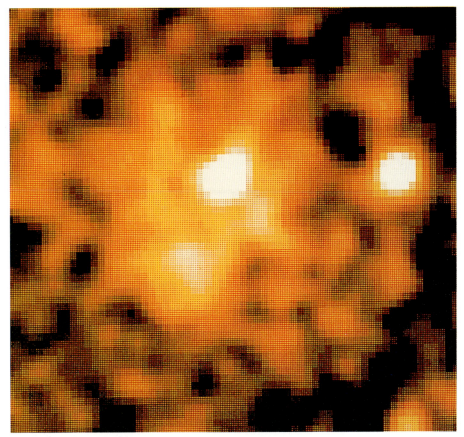

10.31 *X-ray, 3–10 nm, Position Sensitive Proportional Counter, Rosat*

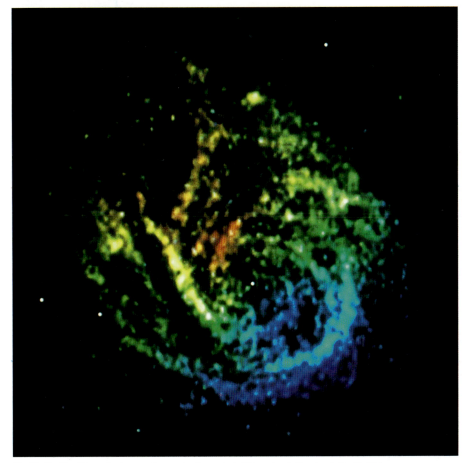

10.32 *Radio, 21 cm hydrogen line, velocity coded, Westerbork Synthesis Radio Telescope*

radio wavelengths. Each is three times the size of the giant nebulae in our Local Group of galaxies, such as the Tarantula Nebula (Fig. 8.24) in the Large Magellanic Cloud or NGC 604 in M33 (Fig. 10.15). Up to a thousand hot massive type O stars within each nebula are heating up ten million solar masses of gas. Put one of these nebulae in place of the Orion Nebula, and it would fill half the night sky with a brilliance to rival the Moon!

Astronomers do not yet know whether these nebulae are scaled-up versions of the nebulae in our Galaxy: a single cluster containing hundreds of O stars and tens of thousands of others, surrounded by a single cloud of gas. Alternatively, they may consist of dozens of more modest nebulae, closely packed as a rash of star-formation spreads along one of M101's spiral arms.

The fury in M101 also shows up at X-ray wavelengths. **Fig. 10.31** covers the same region as the previous three images, and is colour coded for brightness: the most intense regions are white, with fainter parts yellow and orange. The general X-ray glow filling the whole of M101 comes from gas at a temperature of a million degrees. Blasted out by supernovae in its giant starbirth regions, this superhot gas permeates the galaxy and surrounds it with an X-ray emitting halo. Two outlying X-ray sources are probably binary star systems, where a neutron star or black hole is tearing gas from a companion. The brilliant X-ray source right in the middle of M101 may be revealing gas swirling round a massive black hole in the galaxy's heart.

Radio observations can reveal the rotation of gas much further out in the galaxy's disc. In the velocity-coded image of emission from cold hydrogen (**Fig. 10.32**), green and blue show successively higher velocities of approach, while orange and red arms are receding. M101's southern (lower) side is clearly rotating towards us, while the northern (top) side is being carried away. The map also reveals that M101 is actually tipped up with respect to us along a line joining the most intense red and blue regions, and not face-on as it looks at first sight. The galaxy is also rather lopsided, perhaps because it has been disturbed by other galaxies in the loose cluster it inhabits.

Fig. 10.32 covers an area even larger than the ultraviolet image in Fig. 10.28 – a region of space 400 000 light years across which extends to nearly twice the diameter of the optical image. Future generations of stars will be born in this vast, extended disc, and as the site of star formation continues to migrate from the centre, M101's already enormous visible extent will grow further.

10.33 *Optical, 1.2 m Palomar Schmidt Telescope*

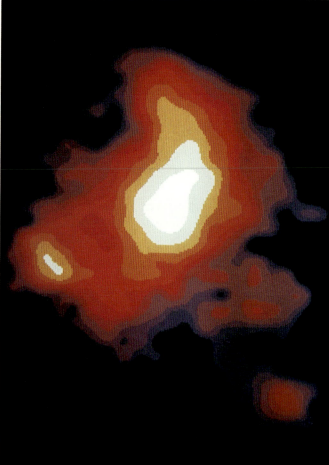

10.34 *Radio, 21 cm hydrogen line, 76 m Lovell Telescope*

M81/82 Group

AT A DISTANCE of 12 million light years, the M81 group of galaxies is one of the nearest small clusters to our Local Group. It covers a smaller region of space (2 million by 1 million light years) but has a similar number of members (thirty-two have so far been identified) and, like the Local Group, is dominated by just a few bright galaxies.

Fig. 10.33 is an optical photograph of the central region of the cluster. (The whole cluster is four times larger, covering an area of Ursa Major more than 350 times that of the full Moon's disc.) M81 (centre) dominates the picture. With a diameter of 100 000 light years, the same as the Milky Way, and a mass of more than 200 thousand million Suns, this spiral galaxy is by far the biggest in the group. Above it in Fig. 10.33 is the peculiar galaxy M82. The other two galaxies are NGC 3077 (left), a small disturbed galaxy, and NGC 2976 (bottom right), a dwarf spiral.

All the galaxies are strongly interacting with one another. Unlike the interaction between the Whirlpool Galaxy and NGC 5195 (Fig. 10.22), this is not obvious at all in visible light. But at radio wavelengths, the picture changes dramatically. Fig. 10.34 shows the distribution of gas throughout the

cluster, colour coded for intensity (and hence amount of gas). White regions are richest in cold gas, while the amounts decrease through grey, orange, shades of red and purple. The plot covers a slightly larger area than Fig. 10.33, and shows clearly that the galaxies are a close-knit group swimming in a huge pool of hydrogen. As well as the gas associated with the individual galaxies, there is a great deal filling the space between them. In fact, this pool accounts for over a quarter of the hydrogen in the group, a mass equivalent to 1400 million suns.

This intracluster gas originated from tidal interactions between the members of the M81 group. Gas has been (and is being) pulled out of the galaxies in streamers, and is slowly diffusing out into the cluster. The main source of gas is M81 itself, as Fig. 10.34 reveals, but the smaller galaxies contribute their share – in particular NGC 3077 to the left. The gas cloud to the lower right of M81, halfway towards the separate blob marking NGC 2976, may be a foreground wisp of hydrogen 'cirrus' in our Galaxy.

The central region of the cluster appears in finer detail in **Fig. 10.35**, an image again made in the emission from cold hydrogen and colour coded in a similar way, with the very faintest regions shown turquoise. The

individual filaments in the pool of gas now show up clearly, along with structures within the individual galaxies. NGC 2976 is out of this field of view, but NGC 3077 lies within the boomerang shape (left) and the cloud of hydrogen at the top surrounds M82.

The spiral shape of M81 is prominent in Fig. 10.35. This finer resolution shows up a 'hole' in the centre of M81, a common feature of Sb galaxies, which have a prominent central bulge and tightly-wound arms. In the more open Sc galaxies, such as M101 (Fig. 10.32), the hydrogen extends right in to the centre.

The inner ring of hydrogen in Fig. 10.35 coincides with the bright optical arms, but the cold hydrogen extends much further out, reaching over halfway to M82 towards the top of the frame. At the bottom, a spiral arm extends out into a 300 000 light year long 'bridge' that forms the bright boomerang to the left of NGC 3077 and then swings up again towards M82.

The hydrogen emission from NGC 3077 itself is very weak: it appears only as the small bright spot within the boomerang. Almost certainly, it was once an irregular galaxy with a plentiful supply of gas. But the tug of M81 has ripped its gas away to

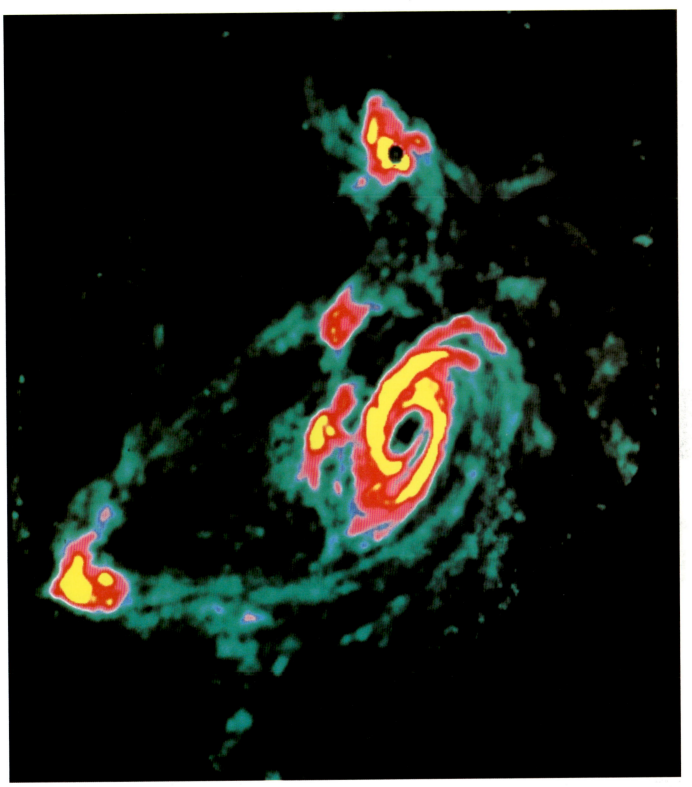

10.35 *Radio, 21 cm hydrogen line, Very Large Array*

form the bright cloud that curves around to the left of the galaxy. NGC 3077 has been severely disrupted, and may end up as a gas-poor elliptical galaxy.

The cloud of hydrogen surrounding M82 has also been distorted. In the inner parts, it matches the orientation of this edge-on galaxy, but the outer reaches of the hydrogen disc are twisted and pulled out into tails that stretch for over 100 000 light years. Undoubtedly, the culprit is again the gravity of massive M81.

The cloud of hydrogen just to the left of M81 provides more evidence for a close encounter between the two major galaxies in the group. It consists of hydrogen torn from M82 as its orbit carried this galaxy through the outer parts of M81's disc, about 200 million years ago. Some stars ripped from M82 have been scattered here to form the backbone of a tiny irregular galaxy. In the cosmic collision, some of the gas has been squeezed into a denser 'molecular cloud' that is poised to begin spawning new stars. From the debris of this galactic crash, a new galaxy is being born.

10.36 *Optical, enhanced true colour, 5 m Hale Telescope*

10.37 *Radio, 21 cm hydrogen line, velocity coded, Westerbork Synthesis Radio Telescope*

M81

'A LITTLE OVAL . . . nebula near the ear of the Great Bear' was how the eighteenth century French astronomer Charles Messier described number 81 on his list of fuzzy patches that could be confused with comets. A moderate-sized telescope reveals a large dim halo surrounding the bright oval centre, but the spiral shape of this galaxy is difficult to discern by eye.

Fig. 10.36 shows why. This optical image was taken through a range of filters to emphasize the subtle colours in M81. The human eye is most sensitive to yellow light from average middle-aged stars like the Sun. These stars form a relatively featureless disc in M81. The spiral arms, on the other hand, are highlighted by bluish radiation from extremely hot stars, which shows up well in photographs.

A radio telescope tuned to the 21-centimetre radiation from hydrogen shows the spiral arms extending much further – and reveals their motion. Fig. 10.37 is colour coded for velocity: red regions have the highest velocities of

recession, blue the highest velocities of approach, while yellow and green lie in between. The spiral arms at the bottom are coming towards us, while the upper portion is turning away. Since the right-hand side of the galaxy lies closer to us, M81 is rotating in an anticlockwise direction with its arms trailing as it spins.

Both M81 and the Andromeda Galaxy are type Sb spirals, with large central bulges, and the 21-centimetre images of each (Figs. 10.4 and 10.37) reveal a central 'hole' where the interstellar hydrogen atoms have disappeared, either in making stars or in the formation of hydrogen molecules. But these observations also reveal a major difference between the 'sister' galaxies. The rotation speed in Andromeda – as in most spirals that have been studied in this way – stays high to large distances from the centre, indicating the gravitational pull of an immense amount of dark matter in and around the galaxy. But the rotation curve of M81 falls off steadily with distance. It is one of the few large spiral galaxies that seems to contain little or no dark matter.

Astronomers divide spiral galaxies into 'flocculent' (patchy) and 'grand design' types. Both the optical and 21-centimetre images show M81 undoubtedly has a grand design: the two main arms can be traced through more than one complete revolution. The stars and gas are piled up into this pattern by the gravitational force of a spiral-shaped 'density wave'. According to theory, a strong external force is needed to set up the density wave in the first place, and the culprit is clearly the companion galaxy M82 (Fig. 10.33).

A density wave should also perturb the oldest stars in a galaxy's disc. Fig. 10.38 is an image of M81 at the red wavelengths emitted by old stars, colour coded for intensity, with the brightest regions white and successively fainter parts orange, pink and blue. A close look shows that the stars in the disc are being shaped into a double spiral by the density wave pattern.

Fig. 10.39 is to exactly the same scale and uses the same colour coding, but it was taken at ultraviolet wavelengths by a telescope flown on the space shuttle, above the Earth's absorbing atmosphere. The

10.38 *Optical, red light, 3.8 m Mayall Telescope*

galaxy is transformed. Now the action is all happening in the spiral arms. At ultraviolet wavelengths, we are seeing the hottest stars, and these outline the spiral arms like streetlights along a road at night.

The density wave does not preferentially bunch hot stars into spiral arms. The process is rather more complex. As interstellar gas crashes into a spiral arm, it is squeezed and begins to form stars of all masses. The massive stars are the brightest and hottest, emitting huge amounts of ultraviolet light. But they live only a comparatively short time – not long enough to wander far from where they were born. So the ultraviolet snapshot pinpoints the stellar birthplaces in M81's spiral arms.

Fig. 10.39 shows 46 major sites where massive stars have just been born. On average, they are three million years old – about the same as the young stars in Orion (Fig. 4.2). The brightest regions in M81, however, shine 100 times more brilliantly than the combined stars of Orion.

The galaxy's bulge accounts for more than half the total brightness of M81 at optical wavelengths, but only 30 per cent in the ultraviolet view. Even so, this is unexpectedly high. All other observations seem to rule out any major amount of starbirth in the centre of M81. Perhaps the old stars here lack the heavy elements that trap short-wavelength radiation, so some ultraviolet is able to seep out.

The centre of M81 is not completely dead, however. **Fig. 10.40** is a view at X-ray wavelengths, covering twice the extent of sky (half a degree square). Within M81 are scattered a dozen individual X-ray sources. They are X-ray binaries where the victim is, in most cases, an old red star – similar to the X-ray sources at the centre of Andromeda (Fig. 10.13). But in M81, the most powerful X-ray source coincides with the galaxy's very heart. Most likely, it is radiation from gas about to disappear down a central massive black hole.

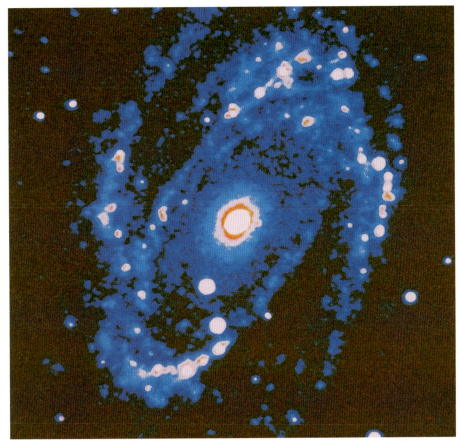

10.39 *Ultraviolet, 249 nm, 38 cm Ultraviolet Imaging Telescope on Astro-1 mission of Space Shuttle Columbia*

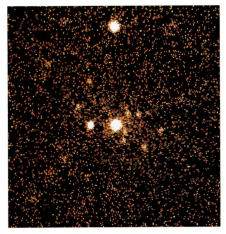

10.40 *X-ray, 0.8–2.5 nm, Position Sensitive Proportional Counter, Rosat*

10.41 *Optical, 665.2 nm, 1 m Jacobus Kapteyn Telescope*

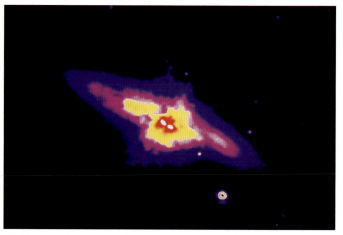

10.42 *Optical, 656.3 nm, short exposure, 1 m Jacobus Kapteyn Telescope*

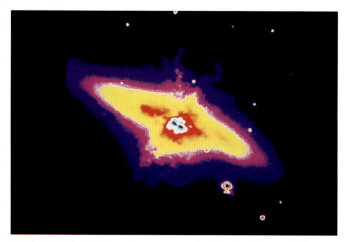

10.43 *Optical, 656.3 nm, long exposure, 1 m Jacobus Kapteyn Telescope*

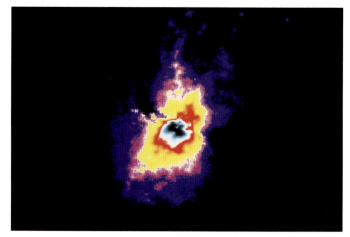

10.44 *Optical, 656.3 nm with 665.2 nm image subtracted to show hydrogen emission, 1 m Jacobus Kapteyn Telescope*

M82

THE GALAXY TO the north of the beautiful spiral M81 has perplexed astronomers for well over a century. Through a small telescope, it is a brilliant sliver of light. But in 1871, Lord Rosse's powerful telescope revealed M82 as 'a most extraordinary object . . . crossed by several dark bands'.

The optical image of M82 shown here in false colour (**Fig. 10.41**) demonstrates the problem of classifying M82. The dimmest parts are coded blue, and brighter regions pink, yellow, red and white. The middle of M82 is broken up into several bright patches. The small dots in Fig. 10.41 are foreground stars in the Milky Way.

The great classifier of galaxies, Edwin Hubble, took M82 as the prototype of 'irregular II' galaxies. His 'irregular I' class were small galaxies whose gas and dust were scattered around in rather a disordered way. Irregular II galaxies were the dustbin of his collection: those he could not classify neatly as irregular I, spiral or elliptical.

Three of the nearest irregular II galaxies lie close to large spirals: NGC 5195, the companion to M51 (Fig. 10.21), and NGC 3077 and M82 in the M81 group (Fig. 10.33). This link was never followed up in Hubble's lifetime, but it is the clue to their strange appearance. In each case, the neighbour's gravitational pull has stirred up clouds of dust that mask the galaxy's true appearance and make it appear 'irregular'.

The blotchy appearance of M82 is highlighted in **Fig. 10.45**, an image at 'far red' optical wavelengths. Here, the deep blue contour delineates the galaxy's fainter outer extent, with regions of increasing brightness pale blue, green, yellow, red, white and intense blue. The superimposed graph shows the brightness as measured from one end of M82 to the other.

While the luminosity of most galaxies increases steadily towards the centre, the graph of M82's brightness goes up and down dramatically. These patches of light and dark are produced solely by streamers of dust – Lord Rosse's 'dark bands' – that cut across in front of the galaxy. The densest band of dust runs just to the left of centre, causing the biggest dip in the graph. Smaller dips reveal at least nine other bands of dust. In places, it cuts down the light from M82 by 98 per cent.

But this is just the beginning of the strange properties of M82. **Fig. 10.42** was taken through a filter that passes not only starlight but also the red light emitted by hot hydrogen. It is colour coded like Fig. 10.41. A thicker region now glows in the middle of the narrow galaxy, centred on two bright spots that peer out like giant cosmic eyes.

A longer exposure through the same filter (**Fig. 10.43**) shows streamers stretching up and down, extending 10 000 light years from the main band of M82. The colour coding is here augmented by black for the brilliant 'eyes' in the centre. It is possible to 'subtract' the starlight – as seen in Fig. 10.41 – and observe the distribution of glowing hydrogen.

Fig. 10.44 is the astounding result. It uses the same colour coding and is to the same scale as the previous three images. This is a giant cloud, glowing in hydrogen light, that appears half as big as M82 itself, yet extends at right angles to the galaxy.

The mystery deepened when astronomers measured the polarization of the light coming from M82. Some astrophysical processes can naturally produce light that vibrates in a single orientation. But the line emission from atoms comes with a completely random mix of orientations, giving no overall polarization. Nonetheless, M82 surprised astronomers once more: the light from the hydrogen-emitting filaments in Fig. 10.44 is polarized.

There is one simple explanation. Light can become polarized when it is reflected: we take advantage of this phenomenon when we use polaroid sunglasses to cut down the glare of the sunlit sea. So M82 has no giant cloud of hot hydrogen. What we see in Fig. 10.44 is a giant halo of tenuous dust particles, which is reflecting hydrogen light from a source hidden in the thick of the galaxy. The shape of the filaments may be showing us simply the direction of individual shafts of illumination from the bright source, just as sunlight can draw bright patterns on a distant landscape as it streams through holes in thick cloud.

Other observations confirm the existence of matter around M82 that can act as a giant cosmic screen. **Fig. 10.46** shows the distribution of dense molecular clouds in M82, colour coded for their speed towards us (blue and green) or away from us (red and yellow). These dust-laden clouds stretch well 'above' and 'below' the line of the galaxy, out to the region of the bright filaments.

The motion of the clouds in Fig. 10.46 provides further evidence for activity within M82. If the galaxy were rotating in a simple way, the velocities would form a simple pattern, from red at one end, through yellow and green to blue at the other end. Clearly, that is not the case. Interstellar matter is moving in or out at a rate of 50 kilometres per second.

The spectacular nature of M82 is brilliantly summed up in **Fig. 10.47** (overleaf), a colour-enhanced optical photograph. This deep computer-processed view, taken through a series of colour filters, emphasises the natural colours of each region of the galaxy and highlights low-contrast details. (The bright red streak at upper right is an artefact.) The underlying galaxy is bluish at the edge and yellow in the middle. A hidden source of hydrogen light is illuminating the surrounding dust cloud (otherwise invisible) in a gaudy display.

But look more carefully. The deep pink patch of reflected hydrogen light is not sitting symmetrically about the centre of M82, but slightly to the right. For many years, astronomers thought M82 was related to 'active galaxies', such as radio galaxies and quasars (Chapter 12), where a massive black hole sits right at the galaxy's heart. Fig. 10.47 demonstrates otherwise. M82's enigmatic powerhouse does not lie at its heart: it is a different kind of erupting galaxy.

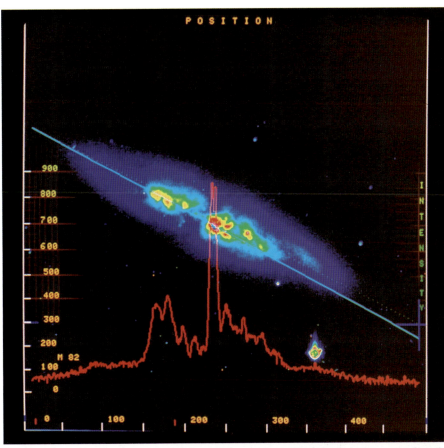

10.45 *Optical, 710–880 nm, 0.6 m reflector, Smithsonian Astrophysical Observatory*

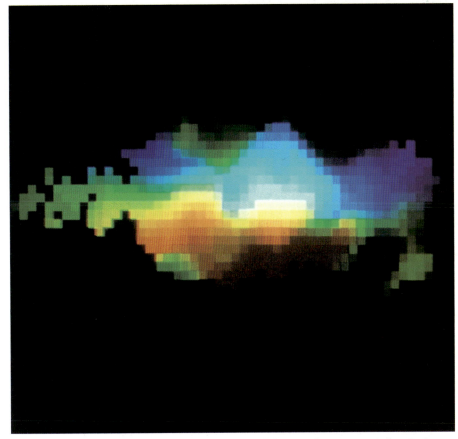

10.46 *Radio, 2.6 mm carbon monoxide line, velocity coded, 13 m telescope, Five College Radio Astronomy Observatory, Massachusetts*

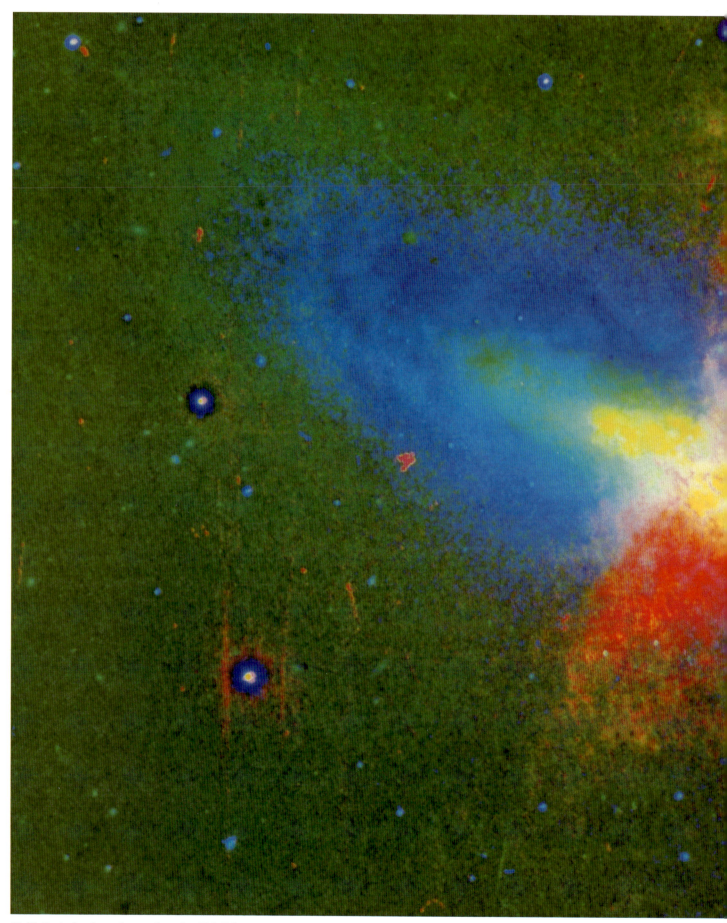

10.47 *Optical, enhanced true colour, 5 m Hale Telescope*

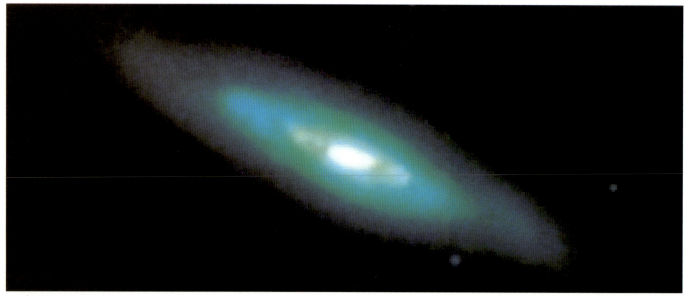

10.48 *Infrared, 1.2 µm (blue), 1.65 µm (green), 2.2 µm (red), 1.3 m reflector, Kitt Peak*

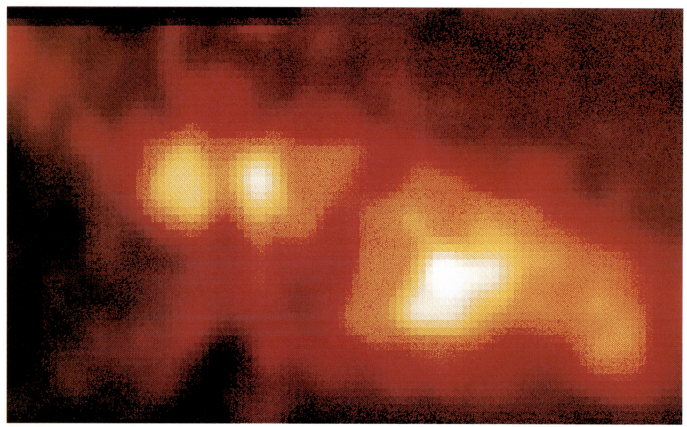

10.49 *Infrared, 12.4 µm, 3 m NASA Infrared Telescope Facility*

Unveiled by infrared astronomy in **Fig. 10.48**, the underlying nature of M82 instantly becomes clear. This image was obtained at wavelengths that cut right through the palls of dust in front of the 'irregular' galaxy, and show it to be as regular as any galaxy could hope to be.

The colour coding is for wavelength, which effectively means temperature, with blue being hottest and green and yellow successively cooler. Stars in the galaxy's disc appear blue, while the central bulge is white. Shining yellow and green are two huge dense clouds of cool gas and dust, 300 light years out from the centre on either side of the bulge.

Fig. 10.48 proves that M82 is in fact a small spiral galaxy, with only one-twentieth the mass of the Milky Way and one-quarter the diameter. It is tilted almost edge on to our line of sight, with the top (North) rim nearer to us.

Longer-wavelength infrared radiation can pin down where the action is.

Fig. 10.49 shows the central region of M82, colour coded for intensity, from red for faint emission, through orange and yellow to white for the brightest parts. It is magnified so that the dust clouds seen to either side of the central bulge in Fig. 10.48 now fill the view: at this wavelength, the clouds shine brilliantly, while the bulge of stars between them is totally invisible.

The emission from these clouds is so intense that at mid and long infrared wavelengths M82 appears ten times

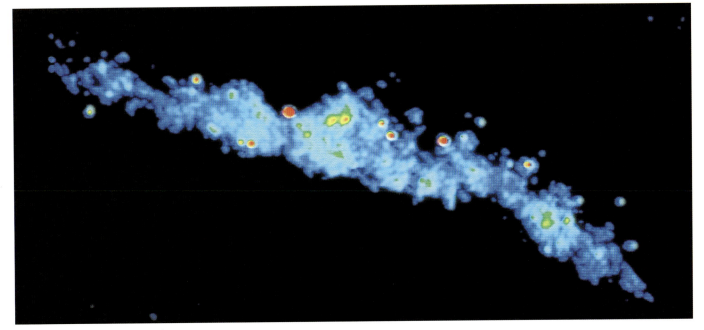

10.50 *Radio, 6 cm, Very Large Array*

brighter than its giant neighbour M81. Despite its distance, M82 ranks equal third in a list of the brightest infrared galaxies in our skies, along with our near neighbour the huge Andromeda Galaxy (and after the two Magellanic Clouds, which are right on our galactic doorstep).

M82 emits 20 times as much infrared as light, a sure sign that the galaxy is a hotbed of star-formation. As a result, it has become the prototype of a new kind of beast in the cosmic menagerie: the 'starburst galaxy'. The Infrared Astronomical Satellite has tracked down hundreds of very distant starburst galaxies, the most extreme generating a thousand times more infrared radiation than M82.

Comparing the optical and infrared views of M82, we find that the brightest cloud in Fig. 10.49 (right of centre) lies exactly at the apex of the filaments shining with hydrogen light (Figs. 10.44 and 10.47). This star-forming cloud is the hidden powerhouse of the galaxy's strange activity. In its centre, hidden from optical view by the thick dust, a massive cluster of half a million hot young stars is heating up its surrounding gas so that the hydrogen shines a brilliant red. This light is streaming out through patches in the dark clouds around, and lighting up a screen of dust in the galaxy's halo.

A radio image reveals the finest details of M82's hidden secrets. In **Fig. 10.50**, the colour coding shows radio brightness, from blue for dim regions through green and yellow to red for the most powerful radio sources. It covers a region twice as wide (2000 light years) as Fig. 10.49. The diffuse radio emission (blue) comes largely from hot hydrogen gas, and near the centre it follows the distribution of dust in the detailed infrared view (Fig. 10.49). The

brighter regions – green and yellow – are the hot nebulae within the dark clouds.

But with starbirth goes stardeath: very massive young stars live only a short while before exploding as supernovae. The radio view also shows M82 as a graveyard of stars. The individual small red sources in Fig. 10.50 are the sites of recent supernovae. In some cases, we are detecting radio emission from the exploding star itself; in others, the supernova remnant created as gases from the explosion crash into dense clouds of interstellar matter.

Some 40 supernovae and supernova remnants appear in Fig. 10.50. All are more powerful than the strongest supernova remnant in our Galaxy, Cassiopeia A (Fig. 6.23), and many are gradually fading. From these data, it seems that supernovae must explode in M82 once every four to five years – compared to a rate of once every fifty years in our own much larger Galaxy.

Over the millennia, the ejecta from these frequent explosions have combined into a giant pool of superhot gas, which shows up clearly at X-ray wavelengths. In **Fig. 10.51**, the faintest X-ray emission is the extensive pale blue area, and successively brighter regions are coded in purple, red, brown and other assorted colours right up to the green patch in the centre. This image is to a smaller scale again: the radio-emitting region in Fig. 10.50 all fits inside the red contour in the X-ray image.

This gas is at a temperature of around 50 million degrees, and it is bursting to escape from the galaxy. Confined by dense dust clouds to either side, it has no choice but to extend 'up' and 'down', in a twin plume of superhot gas.

The inside story of M82 thus has several, interrelated, themes. All can be traced back

ultimately to the gravitational pull of its large companion. M81 ripped up the outermost parts of its small companion, pulling a thick veil of dust across the front of M82 and spoiling its appearance. The gravitational stirring also sent clouds of gas crashing into the centre of M82, and initiated a massive burst of star formation.

This starburst heats up the surrounding dust, to make M82 a brilliant infrared source. When the heavyweight stars reach the end of their short lives, they become powerful sources of radio waves, and fill the galaxy with a pool of ultrahot gas. This gas erupts from either side of the galaxy's disc, leaving gaps where hydrogen light can shine through and illuminate filaments of dust in vivid pink.

So much for the conclusion reached by Edwin Hubble in 1936 – in a rare lapse for this pioneer of galaxy research – when he dismissed the irregular-looking M82 as 'merely nondescript'!

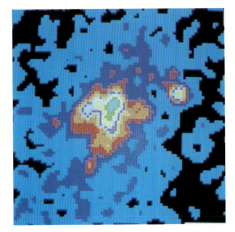

10.51 *X-ray, 0.4–8 nm, High Resolution Imager, Einstein Observatory*

11 x-ray & gamma ray astronomy

JUST BEFORE MIDNIGHT on 18 June 1962, a small Aerobee rocket blasted off from New Mexico under the light of a Moon just one day past full. The rocket soared on an arc which took it 225 kilometres high, well above the atmosphere, before it fell back to Earth. During its few minutes in space, an unusual payload, a trio of Geiger counters, was exposed to the sky.

The flight was supposed to detect X-rays coming from the Moon; but the Geiger counters – the first successful 'X-ray telescopes' – did not 'see' the Moon at all. Nor did they see the bright stars in the sky; nor the band of the Milky Way stretching across the summer heavens. Instead

they saw a dim glow from the entire sky, and even more unexpected, they found a brilliant source of X-rays in the constellation Scorpius – in a position where there is no bright star or radio source.

This flight marked the beginning of X-ray astronomy. The Sun had already been observed in X-rays, but now, for the first time, astronomers could observe the depths of the Universe in the shortest wavelengths of radiation – the X-rays and gamma rays (Fig. 11.1).

There is no hard and fast dividing line between the shortest ultraviolet rays and the longest X-rays. In fact, the extreme ultraviolet (EUV) waves described in Chapter 9 are

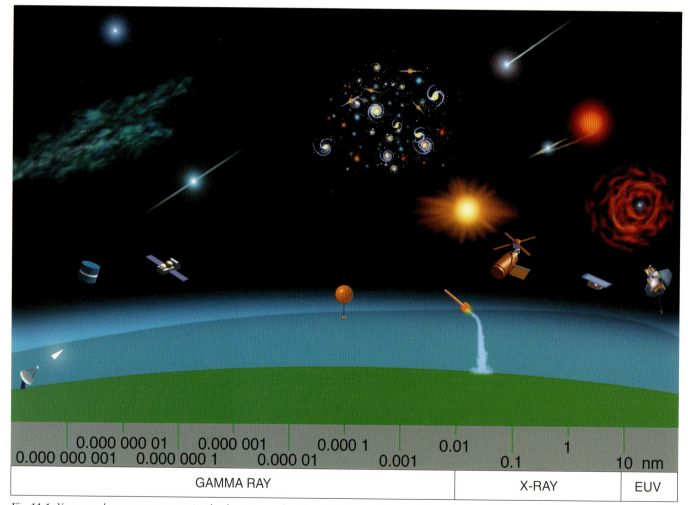

| 0.000 000 01 | 0.000 001 | 0.000 1 | 0.01 | 1 | |
| 0.000 000 001 | 0.000 000 1 | 0.000 01 | 0.001 | 0.1 | 10 nm |

| GAMMA RAY | X-RAY | EUV |

Fig. 11.1 *X-rays and gamma rays comprise the shortest wavelengths of all: the scale here could continue indefinitely to the left, to shorter wavelengths and higher energies. Long-wavelength X-rays abut the extreme ultraviolet (EUV) region. The border between X-rays and gamma rays at 0.01 nanometres (nm) is somewhat arbitrary, but generally speaking X-rays come from extremely hot gas while gamma rays are produced by other processes. All these radiations are absorbed by atoms high in the Earth's atmosphere, but gamma rays can penetrate far enough down to be observed from high-flying balloons. The shortest (highest energy) gamma rays carry enough energy to light up the atmosphere, and the resulting burst of Cerenkov light can be picked up by specially-designed optical telescopes on the ground (far left). X-ray satellites such as Rosat (far right), the Einstein Observatory and the Skylab space station have investigated sources ranging from the Sun's corona, through supernova remnants and black holes in double star systems to clusters of galaxies and jets in distant quasars. The gamma-ray satellites COS-B (far left) and the Compton Gamma Ray Observatory have detected pulsars, clouds of interstellar gas bombarded by high-speed subatomic particles and the cores of distant active galaxies.*

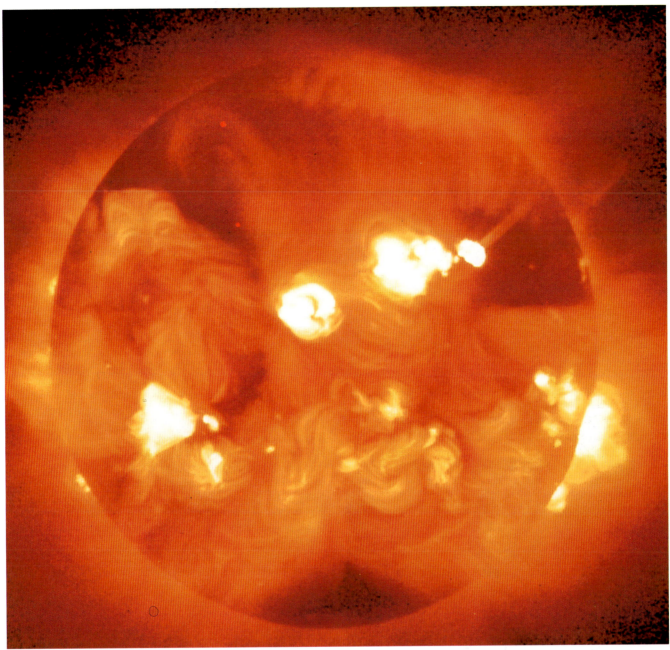

Fig. 11.2 *The Sun is the brightest object in the sky at X-ray wavelengths, and was the first celestial X-ray source to be detected. But a view in X-rays resembles a negative image of the Sun's appearance in ordinary light. Its surface is too cool – at a 'mere' 5800 K – to emit X-rays, and appears here as a dark sphere. The tenuous atmosphere further out, the corona, is at a temperature of millions of degrees and produces copious amounts of X-rays. This gas is marshalled by the Sun's uneven magnetic field. Bright patches lie above the strongly magnetic sunspots, which are dark in optical images. The dimmer regions are coronal holes, where hot gas escapes from the Sun and streams out through the Solar System. This image was obtained by the Japanese Yohkoh satellite, on 12 November 1991, and shows the Sun at wavelengths between 0.3 and 4 nanometres.*

something of a no-man's-land between ultraviolet and X-ray astronomy. For convenience, the boundary between the EUV and X-rays proper is usually taken as being at 10 nanometres – a wavelength about a hundred times the size of an atom. Going down through the X-ray wavelengths, there is again no particular boundary between the shortest X-rays and the longest gamma rays, but astronomers tend to make the distinction at a wavelength of 0.01 nanometres (about one-tenth of an atom's diameter). X-rays are radiation with wavelengths between 0.01 and 10 nanometres; the rest – right down to the shortest detectable wavelengths – are gamma rays.

We usually think of X-rays and gamma rays as extremely penetrating radiation; hospital X-ray machines produce radiation which can penetrate our soft tissues as if they were not there, and gamma rays are used in industry to scan metal sheets in search of cracks and faulty welds. But both kinds of radiation are eventually stopped as they travel through matter. The Earth's atmosphere contains quite enough gas to absorb X-rays and gamma rays from space well before they reach the ground (Fig. 11.1). The longest wavelength X-rays are stopped high in the outermost layers of the Earth's atmosphere, at heights of around 100 kilometres, almost half as high as the lowest satellite orbits. The shorter wavelengths are more penetrating. X-rays similar to those produced in hospital X-ray machines (about 0.02-nanometres wavelength) get down to an altitude of about 40 kilometres; but even the shortest of the gamma rays so far detected – with wavelengths of about 0.000 000 001 nanometres – can penetrate only as far down as 10-kilometres altitude, about the height of Mount Everest and only a quarter of the depth of the atmosphere.

So X-ray and gamma-ray astronomers must fly their telescopes above the bulk of the atmosphere, on rockets or satellites (although high-altitude balloon flights will suffice for some gamma-ray observations). As a result, X-ray astronomy did not begin at all until after the Second World War, which brought the first high-altitude rockets. The German V2 rocket was built as a military weapon, but after the war, captured V2s became a more valuable – and much more successful – weapon in the scientist's struggle to understand the Universe.

Optical observation had already convinced solar astronomers that the Sun's outer atmosphere, the corona, is extremely hot – at a temperature of one to two million degrees K. Such hot gas should emit X-rays. Geiger counters flown on a V2 in 1948 did indeed detect the predicted X-rays from the solar corona. The subsequent brief rocket lights in the 1950s were followed by satellite observatories which could watch the Sun fairly continuously, including the manned Skylab solar observatory and the Japanese Yohkoh satellite. To X-ray detectors the Sun appears quite different. The round disc of the photosphere which is so blindingly bright to human eyes is not hot enough to emit X-rays. It appears as a black globe, surrounded by strands of the million-degree gases which make up the patchy corona (Fig. 11.2).

Despite the early successes of solar X-ray astronomy, most astronomers were pessimistic about detecting X-rays from more distant stars. The early rocket-borne Geiger counters could detect the Sun only because it is so nearby. If the Sun lay as far away as the nearest stars, it would appear so dim in X-rays that no detector of the 1950s or 1960s could have picked up its radiation. A few bold pioneers, however, believed that unpredicted astronomical objects might exist which are dim at optical wavelengths but bright in X-rays. The decisive rocket flight of June 1962 was funded to investigate whether the Moon produced X-rays as a result of the impact of high-speed solar wind particles from the Sun. But the scientists involved were hoping for higher dividends: the discovery of some new, unknown powerful X-ray source far beyond the Solar System, and that is exactly what they found, in the brilliant X-ray source now called Scorpius X-1 (the brightest X-ray object in the constellation Scorpius).

The first X-ray satellite, Uhuru (Fig. 11.3), was launched from Kenya on 12 December 1970, the anniversary of the country's independence day, and *Uhuru* is Swahili for 'freedom'. The American Uhuru and its British, American, Dutch and Japanese successors have located thousands of X-ray sources in the sky. As with Scorpius X-1, such sources do not usually coincide with a bright star or galaxy or powerful radio source. But because satellites can measure their positions fairly precisely, X-ray sources have been matched up with faint optical or radio objects.

Since X-rays are an extremely energetic form of radiation, natural X-ray sources must be excessively powerful or violent types of object. Many are very hot clouds of gas. Just as gas at a temperature of a few thousand degrees (like the Sun's photosphere) emits the comparatively long wavelengths of visible light, while gas of from 10 000 to around 1 000 000 K produces ultraviolet and EUV radiation, so gas that is still hotter shines out in X-rays. The corona of the Sun is one example. Thanks to X-ray astronomy, we now know of many places in the Universe where gas is raised to multi-million degree temperatures. Where a compact star (usually a neutron star) is in orbit around another normal star, gases from the latter can be drawn towards the small star by its gravity. The gas swirls around the compact star in a disc where the individual gas streams spiral inwards as friction robs them of rotational

Fig. 11.3 *The first X-ray astronomy satellite, Uhuru, scanned the sky with a proportional counter behind a collimator (large black rectangle), as it rotated. The optical star- and Sun-sensors (round apertures above) revealed the location of scans on the sky.*

Fig. 11.4 *X-ray image of a black hole: Cygnus X-1 observed by the Rosat observatory. Optical astronomers have found that this powerful X-ray source coincides with a massive star being swung around by an unseen companion. The X-rays are emitted by a stream of gas ripped from the star by its compact and invisible neighbour. In many of these 'X-ray binary systems' the companion is a neutron star, but in Cygnus X-1 its gravitational pull reveals the companion is well over the limit for a neutron star. It is almost certainly a black hole as massive as ten suns.*

energy. The friction also heats up the gas until its temperature is 100 million degrees or more. Scorpius X-1 is a double star system like this, where the gas shines ten thousand times brighter in X-rays than does the Sun at all wavelengths. In some cases, the X-rays come in regular pulses as the neutron star spins round; in others, pent-up gas is suddenly released, to give an X-ray burster. Cygnus X-1 (Fig. 11.4) is an X-ray source where the compact 'star' is too massive to be a neutron star: instead it is a black hole, a collapsed object with such powerful gravity that nothing can escape.

Another type of X-ray source is the supernova remnant. When a star blows itself apart in a supernova explosion, the expanding gases can also reach multi-million degree temperatures. Much farther out in space, we find that clusters of galaxies are embedded in enormous tenuous pools of gas millions of light years across, at a temperature of some 100 million degrees.

X-rays can also come from regions filled with very energetic subatomic particles like electrons, which are either moving through a magnetic field or interacting with longer wavelength radiation. The hot gas streams falling from one star onto another compact star can generate fast electrons which add to the supply of X-rays. In the centres of quasars, gas streams are spiralling into massive black holes; the electrons convert some of their energy into X-rays before they disappear into the black hole.

The early satellites picked up the X-rays from space with a kind of detector which was a slightly more refined version of the simple Geiger counter, the *proportional counter*. It not only detects the presence of an X-ray, but measures its wavelength too. The proportional counter (and other types of X-ray and gamma-ray detector) makes use of the fact that electromagnetic radiation is not just a series of waves,

but also comes in distinct packets of energy, called *photons*. The shorter the wavelength, the higher the energy of each individual photon. So each X-ray photon is much more energetic than a photon of ultraviolet or visible light; and within the X-ray region of the spectrum the photon energy increases as we look at shorter wavelengths.

The proportional counter is a 'chamber' filled with gas – often argon – and containing two wire grids or electrodes, which are kept at a high voltage relative to each other. When an X-ray photon enters the chamber, it collides with several atoms of argon and knocks electrons out of them. The shorter the wavelength of the X-ray the more electrons it produces. An X-ray photon of two-nanometres wavelength would typically knock out some fifteen electrons initially, while for a photon with a wavelength of only one nanometre the number would be around thirty. These negatively charged electrons are repelled from the negatively charged grid, and attracted to the positively charged grid. As they speed through the gas, they knock out additional electrons from the gas atoms, so that an initial bunch of 30 electrons can grow to about 300 000 by the time it hits the positive electrode. This avalanche of electrons constitutes a small electric current, which is picked up and amplified by sensitive electronics attached to the electrode. Whenever an X-ray photon hits the detector, there is a quick burst of electric current, just like the burst which is heard as a 'click' in the uranium prospector's Geiger counter. But in the proportional counter, the strength of the current depends on the number of electrons liberated by the photon, and so it gives a measure of the photon's energy – and hence its wavelength.

The early satellites did not have telescopes to provide a focused view of X-ray sources. Their detectors were open to the sky, but set behind a *collimator*, an array of metal bars or slats, often arranged like a honeycomb, which blinkered their view to a region of the sky around 1° in size. It is very difficult to focus X-rays. No substance can act as a lens to bend X-rays to a focus without absorbing them on the way. Curved bowl-shaped mirrors, of the kind used in optical reflectors and in radio telescopes, are of little use either. An X-ray hitting such a mirror is absorbed rather than reflected.

X-rays are only reflected if they hit a metal surface at a very shallow angle, just grazing it. German physicist Hans Wolter pioneered the design of such *grazing-incidence* X-ray mirrors in the 1950s, when he tried to build an X-ray microscope. His microscope came to naught, but the mirror designs were taken up by Riccardo Giacconi, a pioneer X-ray astronomer involved in the discovery of Scorpius X-1, and they have become the basis of today's X-ray telescopes. The Wolter Type I design for a focusing X-ray mirror (Fig. 11.6) looks nothing like a conventional mirror. It is the polished inside of a slightly tapering metal cylinder. The front half of the cylinder is polished to the shape of a paraboloid; the back half to a rather more steeply tapering hyperboloid. An X-ray entering the front end of the cylinder, near to one edge, strikes the tapering inside surface at a grazing angle, and is reflected so that it travels towards the central axis. After this first reflection, the X-ray hits the more steeply tapering hyperboloidal rear half of the cylinder. Again it is reflected, and travels at a steeper angle towards the central axis of the cylinder. It eventually crosses the axis at a focal point well beyond the cylinder itself. The combination of these two reflections means that the Wolter

Type I mirror forms a perfect image of the X-ray source at the focal point.

Such a grazing-incidence mirror must be used to form an image of an X-ray source, but Wolter telescopes have a major disadvantage: they only collect a small fraction of the radiation falling on the front end of the telescope. In fact, an opaque stop must be mounted in the centre of the tube to block off the majority of the X-rays, which would otherwise travel straight down and 'wash out' the image in the focal plane. The answer, in practice, is to mount a

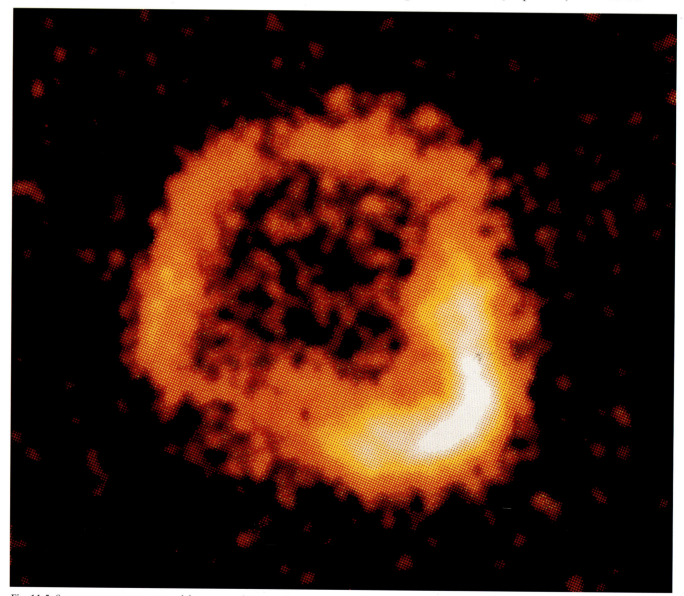

Fig. 11.5 *Supernova remnants are prolific sources of X-rays, though most are faint at optical wavelengths. This bubble of hot gas, RCW 86, lies in the constellation Centaurus. The X-ray view from the Rosat observatory is here colour coded so the fainter regions are orange and the brightest parts white. RCW 86 represents the mortal remains of a supernova that Chinese astronomers saw explode in AD 185. According to a contemporary account 'it was half as large as a mat, showed the five colours and scintillated.'*

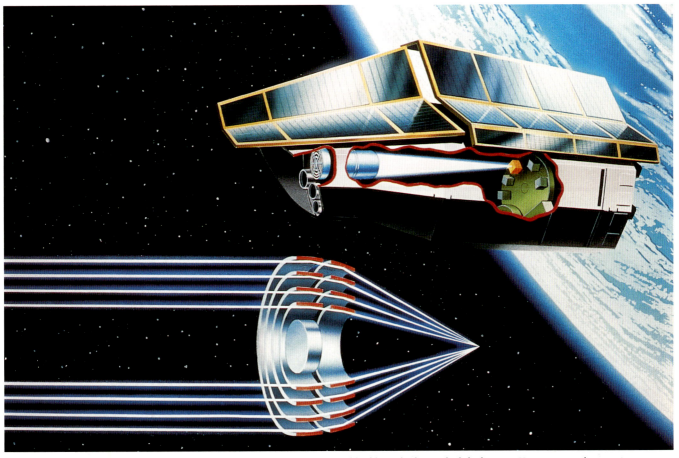

Fig. 11.6 *Cutaway view of the Einstein Observatory shows the mirror assembly (blue cylinder on the left) focusing X-rays onto a detector (orange cube). The inset (shortened left to right) demonstrates how the X-rays are focused by reflection at shallow angles from two successive tapering cylinders; the four nested mirrors have the same focus. The turntable (green) can move any of the detectors (cubes) to this focus. The tubes below the X-ray mirrors are optical star sensors; the solar panels (top) provide power.*

second telescope inside the tube of the first to intercept and focus radiation falling nearer the centre. This smaller, 'nested' telescope is constructed so that it focuses X-rays to exactly the same focal point as the first, and so adds its X-ray image in exact register to the image formed by the larger cylinder.

Giacconi's first X-ray telescope formed part of the solar observatory – the Apollo Telescope Mount – on the Skylab space station (Fig. 9.7). The focused X-ray images were recorded on film which was returned to Earth by the Skylab astronauts, and the exquisite views of the Sun's corona proved that an X-ray telescope could produce pictures with detail as fine as that from optical telescopes.

In order to look at the faint, distant sources beyond the Solar System, X-ray telescopes must be much larger and heavier than the simple proportional counters mounted on the early generation of X-ray satellites. The chance to fly one came with the series of three huge High Energy Astrophysics Observatories (HEAOs) in the late 1970s. These satellites each weighed some 3 tonnes, and were 5.8 metres long – a HEAO satellite standing on end would reach as high as a house. The second HEAO was built around an imaging X-ray telescope. This epoch-making satellite was renamed the Einstein Observatory, to honour the great physicist Albert Einstein whose centenary occurred in 1979.

The Einstein Observatory's telescope (Fig. 11.6) consisted of four Wolter Type I mirrors nested inside one another,

with the outermost mirror 58 centimetres across. Giacconi's team provided it with four different kinds of X-ray detector. They were mounted on a turntable which rotated to bring any one of them into the position of the focused X-ray image. The Einstein Observatory was thus as flexible as a ground-based optical observatory, where astronomers regularly change the light-detectors or spectrographs according to the kind of observation they want to make.

The Imaging Proportional Counter (IPC) worked in essentially the same way as the proportional counters used in early satellites, except that it measured not only the energy (wavelength) of the incoming X-ray photon, but also the position at which it hit the detector, thus building up a picture of the source. The Einstein IPC was very similar to the human eye in its resolution of detail and in its ability to see different wavelengths – 'colours' of radiation – albeit at wavelengths a thousand time shorter.

The IPC had a simple way of measuring the position of incoming photons – or, rather, the bursts of electrons they produced. Beneath its wide front plastic window were three grids of wire, one below the other. The middle grid was positively charged, the upper and lower grids were negative. When an X-ray photon passed through the window into the gas-filled IPC, it created a burst of electrons which headed for the nearest point of the central grid. On its arrival, the pulse of electric charge produced an electric current there, and also induced pulses of current in the upper and lower grids. These grids were each formed of a positions of thousands of X-ray emitting binary stars in our Galaxy.

Looking beyond the Milky Way, the Einstein Observatory and Rosat see a view very different from the familiar optical photographs. We are used to seeing the Andromeda Galaxy as a spiral-shaped, uniformly glowing expanse of stars too faint to be resolved individually. The X-ray telescope on Rosat sees the Andromeda Galaxy as a loose cluster of about a hundred individual bright spots (Fig. 10.13). These are the immensely powerful X-ray sources caused by gas streams falling from a star onto a compact companion star, like Scorpius X-1 in our own Galaxy.

Farther out in space we come across active galaxies – Seyfert galaxies, radio galaxies and quasars – and in these the Einstein Observatory's telescope picked up powerful X-ray emission from the very centres. In some galaxies, the X-ray output fluctuated dramatically in a matter of a few hours, indicating that the gas clouds producing the radiation are very small. This has strengthened most astronomers' conviction that a quasar consists of gas clouds held tightly around a central supermassive black hole. Rosat has detected over 20 000 active galaxies emitting X-rays.

Rosat has also made a deep study of the background of X-rays that has been a mystery since it was discovered in that first rocket flight in 1962. At least half of this radiation, it turns out, comes from individual very distant quasars and starburst galaxies. In all likelihood, the rest comes from similar objects at even greater distances, their individual contributions melding to make the sky look more-or-less uniformly bright in X-rays.

And the new X-ray instruments have provided strong

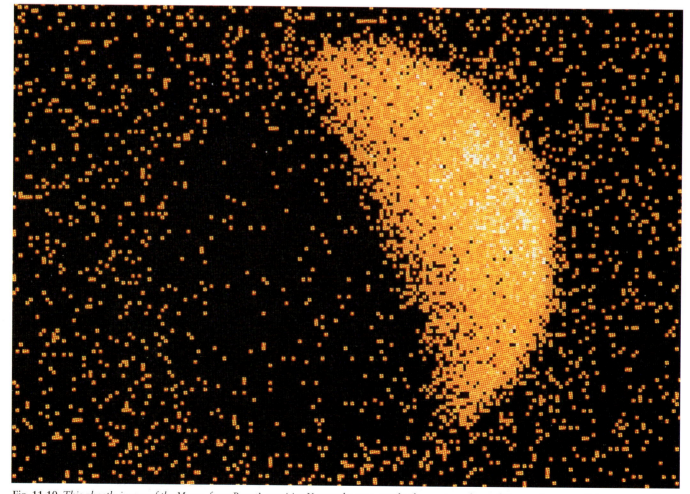

Fig. 11.10 *This ghostly image of the Moon, from Rosat's sensitive X-ray telescope, marks the coming of age of X-ray astronomy. Thirty years earlier, the first X-ray experiment had been launched to detect X-rays from the Moon: it failed because its Geiger counter was too insensitive, but did pick out an unexpected background of X-rays coming from all over the Universe. The X-ray background shows up in this Rosat image as a random scattering of dots, added to the dots representing electronic 'noise' in its detectors. Note that the dark side of the Moon is actually silhouetted against the glowing background – an impossible sight at optical wavelengths.*

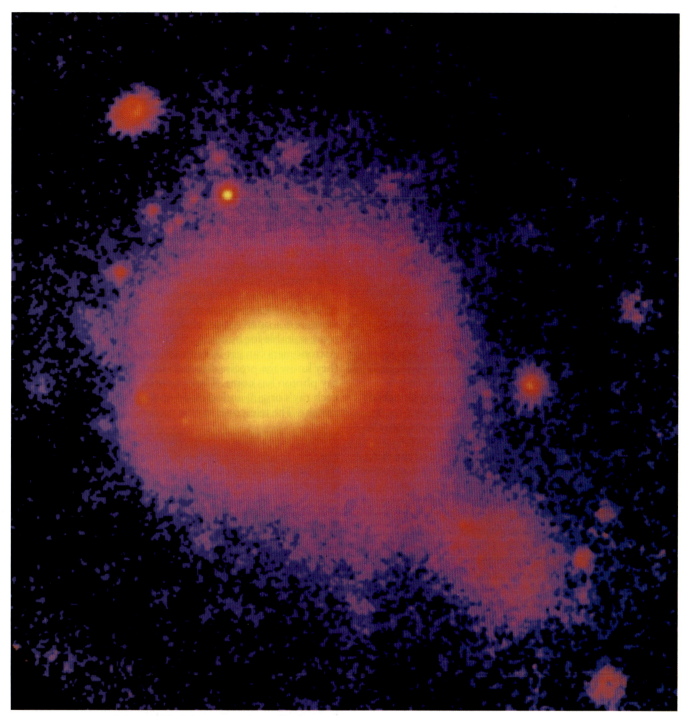

Fig. 11.11 *Rosat reveals hot gas filling the centre of the Coma cluster of galaxies, some 300 million light years away. Colour coding shows X-ray brightness, running from yellow for the most intense region through orange and red to purple for the faintest parts. Small outlying blobs show gas trapped in individual galaxies, but in the centre the hot gas has been stripped from galaxies to form a communal pool. The spur to the lower right is a smaller cloud of gas surrounding the galaxy NGC 4848: eventually it will merge into the larger central pool. X-ray astronomy can hence give astronomers invaluable clues to the evolution of clusters of galaxies.*

evidence that familiar matter – made of protons, neutrons and electrons – is only a minor constituent of the Universe. The Einstein Observatory could make out the pools of hot gas that sit within rich clusters of galaxies; Rosat has been able to observe them in detail. The total gravitational pull on this gas turns out to be far greater than the pull expected from the galaxies alone. This discovery supports other hints that 90 per cent of the Universe consists of *dark matter*. Dark matter does not produce light or any other radiation, and is detectable only by its gravitational influence: it may

consist of some different kind of subatomic particle that does not interact with ordinary matter. Rosat has found that much of the dark matter lies not in large galaxy clusters, but in minor clusters like our own Local Group of galaxies.

The multinational nature of Rosat is typical of X-ray astronomy: it is a field where many nations have been in the forefront. One forthcoming mission, Spectrum-X, is a Russian satellite carrying a pair of large European-built telescopes. And the European Space Agency is planning a one degree – in optical terms, that would be just sufficient to distinguish the stars of Orion's belt from one another.

But there are techniques that can probe even finer details of the gamma-ray sky. The front-runner is the *coded-mask telescope*. Prototypes have been flown on balloons, the Russian manned space station Mir and a Russian-French satellite. This satellite, Granat, pinned down the position of a bright gamma-ray source near the centre of the Milky Way with a precision of a few arcminutes, proving – to the surprise of many astronomers – that it did not coincide with the galactic centre itself.

A coded-mask instrument looks more like a conventional telescope, with a detector at the lower end that can identify the position of each incoming photon: it may be either a scintillator or a spark chamber. At the top, instead of a mirror or lens is a mask made of small blocks of lead or tungsten, which absorb incoming gamma rays. The blocks are arranged in a random pattern, looking like the black squares in a crossword puzzle.

As a gamma-ray source 'shines' on the telescope, it casts a shadow of the mask on the detector. By measuring the position of the shadow accurately, astronomers can work out precisely where in the sky the source lies. If there are two sources in the field of view, there will be two overlapping shadows. Because every part of the mask is different from every other, computer techniques can distinguish the two shadows and work out the position of both sources. In fact, it is possible to work out the positions of all gamma-ray sources in the field of view, and produce a map of the sky with a resolution of around one arcminute – about as good as the human eye.

While some researchers are tackling the problems of resolution, other astronomers are busy pushing back the high-energy frontier of the gamma-ray spectrum. Here they face the ultimate problem with the pre-packaged nature of radiation. The photons individually carry so much energy that there are only a small number of them. As a result, you need a large collector in order to pick up any photons at all.

Some astronomers have tackled this problem by using the biggest scintillation detector we have to hand – the Earth's atmosphere. When a gamma ray is absorbed in the air, it produces a flash of light, and the flash from the highest energy gamma rays can be picked up by an ordinary optical telescope equipped with an electronic light detector. The telescope needs a large mirror to see the faint flashes, but this mirror does not need to be accurately made. Such telescopes have picked up flashes that were produced by gamma rays with an energy of a billion MeV – the shortest radiation so far detected from space.

No gas cloud can be hot enough to generate such short wavelength radiation in the way that stars produce light and cool dust clouds shine in the infrared. Gamma rays come indirectly from the extremely fast subatomic particles generated in the most violent spots in the Universe:

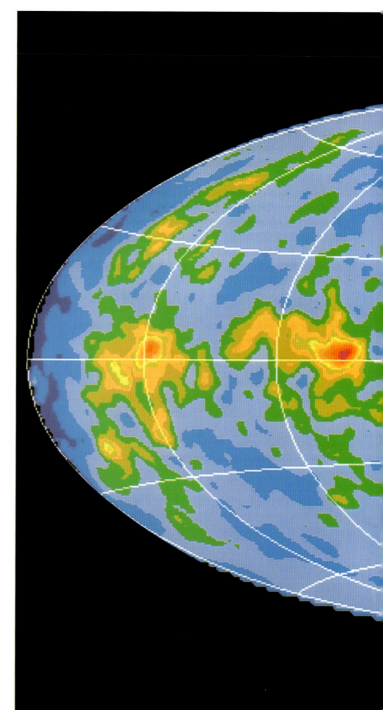

supernova explosions, the powerful magnetic fields around pulsars, the gravitational wells around huge black holes in the centres of active galaxies – and perhaps from other, unknown regions of cosmic activity which gamma-ray detectors may eventually reveal to us.

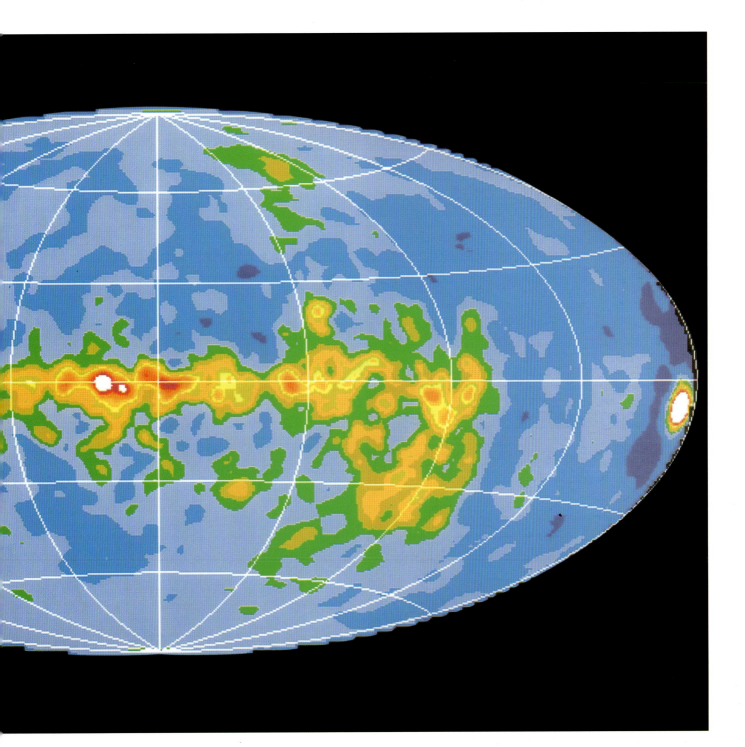

Fig. 11.14 *First view of the whole sky at middle gamma-ray wavelengths reveals some old friends – and some surprises. This image from the Comptel instrument on the Compton Gamma Ray Observatory is colour coded for intensity, with the brightest parts white, and successively fainter regions red, shades of orange, green, lilac and blue. The Milky Way runs horizontally across the centre. The bright source at extreme right is the Crab Pulsar, while the Vela Pulsar lies halfway in from the right to the centre. The Galactic Centre itself is a fairly bright emitter (red region in middle of image), but the much stronger source (white) just to its left is still unidentified. The source halfway out to the left (red) is the black hole candidate Cygnus X-1.*

12 active galaxies – and beyond

UNTIL THE ADVENT of the new astronomy, scientists regarded the Universe beyond our Milky Way as peaceful and serene, populated by gently swirling spiral galaxies and boring old ellipticals. Cosmology, including the great questions concerning the origin and fate of the Universe, were the province of the philosopher. During the last half century, however, astronomers have spearheaded one of the great cultural revolutions of modern times. It began with violent galaxies and has led to the greatest of all explosions, the Big Bang.

Radio astronomers were in the vanguard. In the late 1940s and the 1950s, they found that some of the strongest radio sources are galaxies billions of light years away. It's as if the visible night sky were dominated not by nearby stars like Sirius and Betelgeuse, but by objects a thousand times further off than the Andromeda Galaxy.

The radio sky is vividly demonstrated in the composite image (**Fig. 12.1**), behind a photograph of the radio observatory at Green Bank in West Virginia. The radio view was obtained by the 90-metre dish (right), which subsequently collapsed from metal fatigue. The larger objects in the sky are nebulae and supernova remnants in our Galaxy. The small sources are not stars: they are powerful galaxies lying a billion times further away, towards the edge of the observable Universe.

These radio galaxies emit up to a million times more radio power than our Milky Way. And, even stranger, a more detailed look reveals that the radiation is usually generated in a pair of giant lobes, located to either side of the galaxy itself. Energy travels to these radio lobes in the form of a jet or beam of electrons, squirted out from a tiny powerhouse at the galaxy's centre.

Similar active galactic nuclei characterise several other kinds of violent galaxy. The most important type were tracked down in the early 1960s, when optical astronomers identified a handful of radio sources as blue 'stars'. The clumsy name 'quasi-stellar radio sources' was soon shortened to quasars. Although quasars lie billions of light years away, they are among the brightest objects in the sky when observed in most regions of the spectrum, from radio waves through to X-rays and gamma rays. If astronomy had started at these wavelengths instead of visible light, active galactic nuclei would be as familiar as stars and nebulae.

Quasars and other active galactic nuclei pack the power of a hundred galaxies into a volume no larger than the Solar System. Most astronomers explain them as a disc of hot gas which is circling a very massive black hole right at the galaxy's heart. As the gas gradually spirals down into the black hole, its orbital energy is transformed into radiation of all kinds, and into the speeding electron jets.

In fact, the way we classify an active galaxy may simply depend on how we view it. The key to this theory is a doughnut of dark dust surrounding the bright accretion disc. If we view it edge-on, the doughnut hides the bright disc, and we have a radio galaxy. If we happen to be looking *into* the doughnut, we observe the bright centre and classify the object as a quasar.

The advent of X-ray astronomy in the 1960s and 1970s brought a new spate of quasar identifications. And X-ray satellites also pinpointed another unexpected component of the Universe. Clusters of galaxies are filled with pools of hot gas, at a temperature of typically 100 million K. Invisible to optical telescopes, this superhot material shines brilliantly at X-ray wavelengths.

And over the immense reaches of space now accessible to telescopes, light and other radiation can be subjected to strange tricks. The gravitational pull of a massive galaxy or a mighty cluster of galaxies can focus the radiation passing by. These gravitational lenses can distort the image of more distant galaxies and quasars into weird mirages: split images, banana shapes and complete rings.

Radio telescopes can reach furthest of all. Perhaps the greatest astronomical breakthrough this century was a faint 'hiss' from the sky, discovered by radio astronomers in 1965. It turned out to be radiation from the Big Bang. Observing this radiation in detail in 1992, the COBE satellite found signs of galaxies being born in the fireball erupting from the Big Bang. Armed with the tools of the new astronomy, scientists are beginning to understand both when, and how, the Universe began.

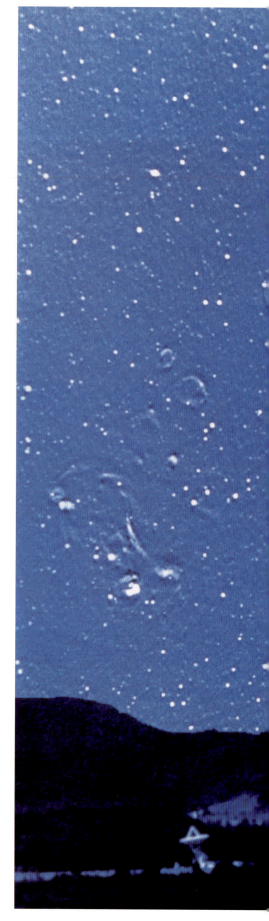

12.1 *Radio, 6.2 cm, National Radio Astronomy Observatory 90 m telescope, Green Bank*

12.2 *Optical, true colour, 3.9 m Anglo-Australian Telescope*

Centaurus A

THE NEAREST ACTIVE galaxy is surprisingly close to us on the extragalactic scale. At 16 million light years, Centaurus A is only seven times further away than the Andromeda Galaxy. It is about the same distance as the Whirlpool galaxy M51; and less than one third the distance of the better-known radio galaxy M87 in Virgo. It is the third strongest radio source in the sky (after Cassiopeia A and Cygnus A), and a source of X-rays and gamma rays. But Centaurus A lies a long way South in the sky, and is always below the horizon (or rises to a very low altitude) as viewed from the traditional optical and radio observatories of Europe and North America. Only in the past couple of decades, with the opening of major new observatories in Australia and Chile and satellites which can observe the whole sky, has Centaurus A been given the attention it deserves.

The true-light photograph (**Fig. 12.2**) shows the central concentration of stars in Centaurus A, crossed by an unusual band of dust which obscures the stars behind it. This ball of stars is 6 arcminutes in apparent size, about 30 000 light years in actual extent, and contains almost a million million stars. The combined light from these stars makes Centaurus A one of the brightest galaxies as seen from Earth: it appears in the background of many

photographs of Halley's Comet taken in 1986 (Fig. 2.35). At magnitude 7, Centaurus A is easily seen in binoculars.

British astronomer John Herschel noticed the galaxy's unusual appearance when observing from South Africa in the 1830s: 'a most wonderful object . . . cut asunder . . . by a broad obscure band'. It was listed in the New General Catalogue of 1888 as NGC 5128. The galaxy was relatively little studied until 1949, when pioneering radio astronomers in Australia identified NGC 5128 with a radio source discovered the previous year and named Centaurus A – the strongest radio source in the constellation Centaurus. (It was among the first three radio sources to be identified, the others being Virgo A as the galaxy M87 and Taurus A as the Crab Nebula.)

New optical techniques have revealed that the stars and dust seen in ordinary photographs like Fig. 12.2 form only the central regions of a very much larger galaxy. David Malin has brought out the faintest extremities of Centaurus A in **Fig. 12.3**, where the inset, showing the familiar central regions, is to the same scale! The original photograph was taken with the 1.2-metre UK Schmidt Telescope in Australia, and the inset is a normal print from the negative. The lightly exposed image of the galaxy's faint extensions is recorded on the negative, but resides in the surface grains of the emulsion. On ordinary

printing it is lost in a background caused by chemical 'fog' scattered throughout the emulsion's thickness. Malin's technique of 'photographic amplification', however, prints only the image in the surface grains, producing Fig. 12.3 from the same negative.

As seen here, the main body of the galaxy is 150 000 light years by 130 000 light years in extent, with an apparent size (33 by 27 arcminutes) as large as the Moon. In three dimensions, Centaurus A is a rugby football shape (a prolate spheroid). From either end, faint streamers of stars reach out from the galaxy's body to double its total extent to a third of a million light years.

Further 'photographic amplification' brings up extremely faint swathes of light which cover the entire picture; some of the brighter ones are seen across the bottom of Fig. 12.3. They are not related to Centaurus A, but are clouds of dust in our own Galaxy. This 'interstellar cirrus' has been raised a few hundred light years from the Milky Way's disc by supernova explosions, and is illuminated by the glow of stars in our Galaxy.

12.3 *Optical, photographically amplified (inset printed normally), 1.2 m UK Schmidt Telescope*

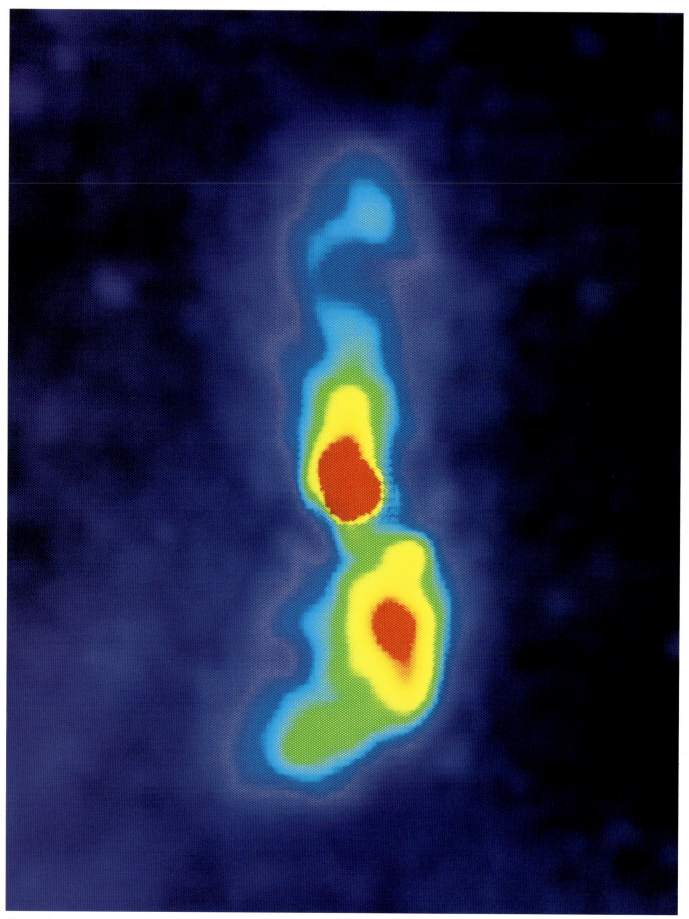

12.4 *Radio, 13 cm, 26 m Hartebeesthoek Radio Telescope, South Africa*

12.5 *Combined wavelengths: radio, 6 cm, Very Large Array; optical, 3.9 m Anglo-Australian Telescope*

If the optical extent of Centaurus A looks large, then its radio structure is absolutely immense. It comprises two huge 'lobes', seen clearly in the radio image (**Fig. 12.4**). The colour coding runs from purple for the faintest regions through blue, green and yellow to red for the most intense central region, coinciding with the galaxy itself. The radio lobes stretch over 9° of the sky – almost twenty Moon breadths, or one-tenth the way from the horizon to overhead, even though Centaurus A lies 16 million light years away.

In real terms, the lobes cover 2½ million light years, a distance further than the separation of the Andromeda Galaxy and our Galaxy. Centaurus A is, in fact, one of the largest radio galaxies known. To compare with optical photographs, the faintest regions in the photographically amplified view (Fig. 12.3) cover only the central one-tenth of the radio span.

Although Centaurus A has unusually large lobes for a radio galaxy, its radio output is quite weak – about a thousand times stronger than the output of a spiral like the Andromeda Galaxy, but representing only one-thousandth of the light output from the galaxy's stars.

The lobes are produced by flows of fast electrons heading North and South from the central galaxy. These electrons generate two elongated clouds of magnetic field, which in turn trap the electrons. As the electrons whirl back and forth through the clouds, they produce synchrotron radiation in the form of radio waves.

The electrons now in the outermost part of the radio-emitting lobes were shot out from the centre about 100 million years ago. Since then, the flow of electrons has changed in direction, swinging 40° anticlockwise so that the most recent, innermost portions of the clouds lie along a line roughly 'five o'clock-eleven o'clock'. Their strength has also varied. The flow of electrons to the South apparently ceased altogether for a while, leaving a gap in the radio emission between the galaxy and the bright spot at 'five o'clock'. During this period, the flow to the North was stronger, elongating the bright region surrounding the galaxy towards 'eleven o'clock'.

Fig. 12.5 reveals the most recent burst of activity in Centaurus A. The colour-coded radio image – the central bright (red) region of Fig. 12.4 magnified sixty times – is superimposed on an optical photograph. (The strange blue star in the dust band is a supernova that exploded in 1986, and was recorded on only one of three black-and-white plates taken to make this composite – the photograph taken through a blue filter!) In the radio image, colours represent intensity, with the brightest regions red, and less intense regions yellow, green and blue. Most of the emission comes from the lobes at the top left and lower right, lying about 20 000 and 10 000 light years out.

The flow of electrons towards the top left lobe is revealed as a long thin 'jet' (blue and green), shooting outwards from the centre of Centaurus A. The source of the electrons – the galaxy's active nucleus – appears as a small bright (red) blob on the radio image, but it is completely hidden from optical telescopes by the band of dust running in front.

While the long wavelengths of radio waves first alerted astronomers to the strange nature of Centaurus A, the very shortest wavelengths are crucial to understanding the beast that lurks at its centre.

Fig. 12.6 is an X-ray view of the centre of Centaurus A, covering a region 25 000 light years across. Dim regions are blue and green, with brighter parts of the image red and white. Striking diagonally across is the jet of electrons, already seen in the radio image (Fig. 12.5). As the electrons move outwards from the nucleus, they produce radiation over a range of wavelengths. The most energetic electrons lose their energy fastest, so the high-energy X-ray jet is brightest near the centre of the galaxy. It fades towards the upper left, where the radio lobe begins.

Most of the galaxy's X-radiation, however, comes from the tiny nucleus at the centre of Centaurus A. It shows as the central white spot in Fig. 12.6, but the size it appears here is determined by the resolution of the X-ray telescope. In reality, it is far smaller. In 1973, the X-rays from Centaurus A brightened in just a few days, implying that the core is only a few 'light days' across – about one-hundredth of a light year.

But this is just the beginning of a high-energy bonanza. **Fig. 12.7** is a scan of a huge region of sky at gamma-ray wavelengths. It is 60° square, covering the four large constellations of Virgo, Libra, Lupus and Centaurus. The colour coding shows successively brighter regions in shades of blue, red and yellow. Most of the structure in this pioneering image (including the apparent source at far right) is just 'noise': the only object actually emitting gamma rays (lower right) is Centaurus A.

Its output of gamma rays is prodigious: 10 000 million times more energy than the Sun radiates at all wavelengths. Centaurus A emits twice as much power in X-rays as at radio wavelengths, and a hundred times more power still in gamma rays. Despite its classification – for historical reasons – as a 'radio galaxy', Centaurus A would be better described as a 'gamma-ray galaxy'!

Fig. 12.7 has very poor resolution, so it blurs Centaurus A enormously. The strength of the gamma rays varies over a few months, so the actual gamma-ray source in Centaurus A must be less than a light year across – and is probably much smaller still.

This gamma-ray source represents such an intense concentration of energy that most astronomers believe it must be the inner part of a hot gas disc surrounding a massive black hole. The X-rays come from less energetic regions further out. The disc can channel electrons into jets beaming out from either side: at present, only the jet heading to the upper left is flowing.

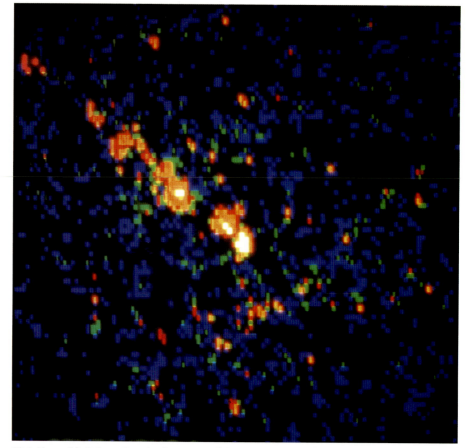

12.6 *X-ray, 0.3–2.5 nm, Imaging Proportional Counter, Einstein Observatory*

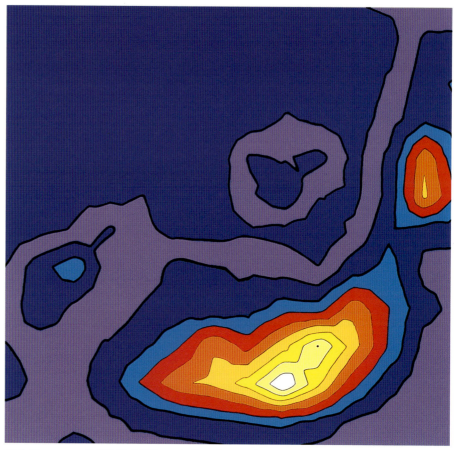

12.7 *Gamma ray, 0.000 06–0.002 nm, balloon-borne telescope*

12.8 *Infrared, 1.2 μm (blue), 1.6 μm (green, 2.2 μm (red), 3.9 m Anglo-Australian Telescope*

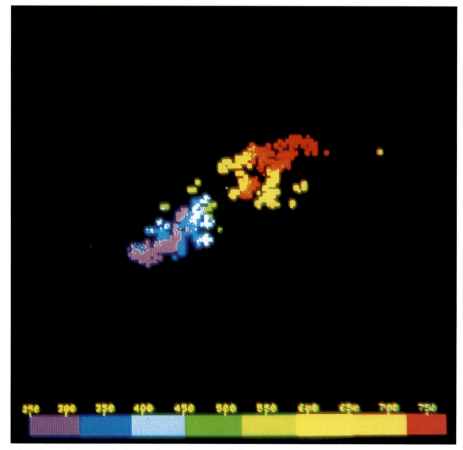

12.9 *Optical, 656 nm hydrogen line, velocity coded, 3.9 m Anglo-Australian Telescope*

The energetic activity in Centaurus A is presumably related to its strange overall appearance: in particular, the band of dust that distorts its appearance. Infrared wavelengths can cut through most of the dust and reveal the galaxy's basic form.

Fig. 12.8 is a near-infrared view of Centaurus A, colour coded for wavelength so that we see the galaxy at wavelengths four times longer than visible light. It covers the central 25 000 light years of the galaxy, the width of the ball of stars seen in the optical photograph (Fig. 12.2).

The region between the two parallel bands of dust in Fig. 12.8 is completely obscured at optical wavelengths, but is relatively transparent in the infrared. The packing of stars increases steadily towards the centre, and it is easy to pick out precisely the location of the galaxy's core. (At optical wavelengths, it is dimmed over a million times.)

The brightness in Fig. 12.8 increases towards the centre in exactly the way expected for an elliptical galaxy. But these galaxies have little by way of gas or dust – so whence the great dust lane of Centaurus A?

In Fig. 12.8, the bright infrared clumps in the densest dust lanes are regions where stars are forming. Towards its outer edge, young blue stars and reddish nebulae can just be picked out in the optical photograph (Fig. 12.2). The light from the nebulae, in particular, can show up the motion of this material.

Fig. 12.9 is an observation of Centaurus A, to the same scale as Fig. 12.2, made with an imaging spectrograph that picks out the red light from hydrogen in nebulae. At each point, the instrument measures the precise wavelength of the hydrogen emission, to reveal any slight changes caused by the motion of the gas (by the Doppler effect). The colour coding shows the deduced speed of the gas, towards us (purple and blue) and away from us (yellow and red). The maximum speed is 250 kilometres per second.

Despite the jumbled appearance of the dust lanes, Fig. 12.9 shows the interstellar matter is orbiting the centre of Centaurus A in a very orderly way. The most likely explanation is that a small spiral galaxy ran into the giant elliptical some 100 million years ago. The interloper's stars have become scattered among those of the giant elliptical. The interstellar matter in the spiral galaxy, however, preserved its original sense of rotation and is now forming a disc around its surrogate galactic centre.

In the stirring that accompanied the merger of the two galaxies, some of the spiral galaxy's dust and gas clouds fell to the centre of the giant elliptical. Here they encountered a massive black hole, and formed into a small energetic accretion disc that is powering the galaxy's great outburst of energy.

12.10 *Optical, 3 m reflector, Lick Observatory*

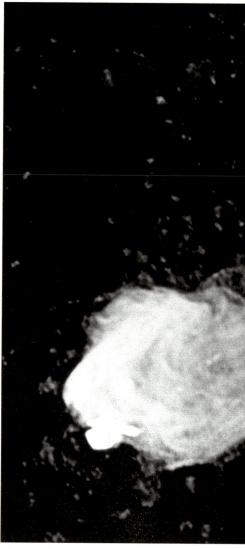

12.12 *Radio, 6 cm, Very Large Array*

Cygnus A

THE SECOND BRIGHTEST radio source in our skies is a galaxy so distant that it appears extremely faint to optical telescopes – at magnitude 15, less than a thousandth as bright as the dimmest stars visible to the naked eye. In this optical photograph (**Fig. 12.10**), the central regions of the galaxy appear as the small dumb-bell shape at the centre: the whole picture is only 2½ arcminutes across, less than one-tenth the apparent diameter of the Moon. Its discoverer Walter Baade was convinced that Cygnus A was a pair of galaxies in collision, and he placed a bet – a bottle of whisky – with his colleague Rudolph Minkowski that the spectrum would show light from hot gas produced in the collision. The spectrum turned out as he predicted and Baade won his whisky – but in the long run the interpretation proved to be wrong. Cygnus A is a single galaxy, with an active quasar core which heats up gas in the centre. The peculiar double appearance in Fig. 12.10 is caused by a band of dust

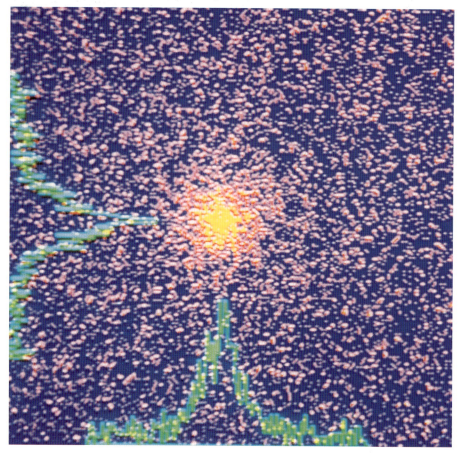

12.11 *X-ray, 0.4–8 nm, High Resolution Imager, Einstein Observatory*

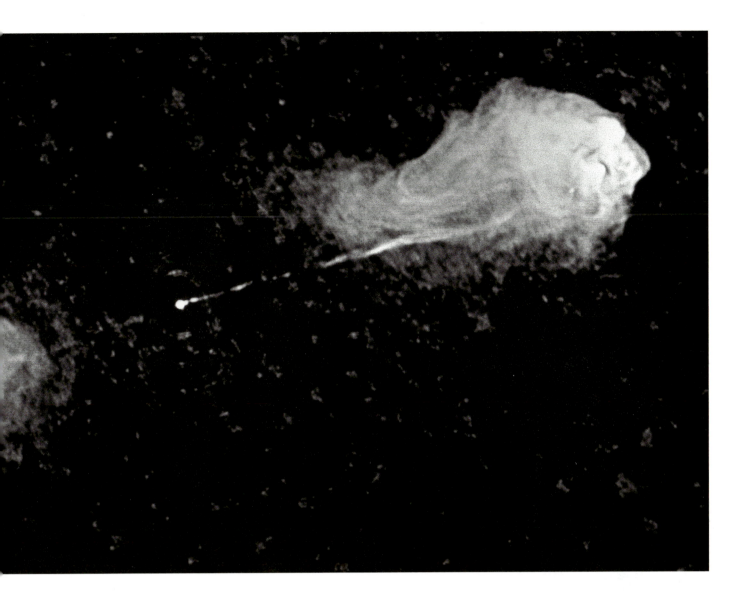

running across the galaxy, like the dust lane in Centaurus A (Fig. 12.2).

The Cygnus A galaxy appears small and indistinct because it lies fifty times further away than Centaurus A, at 740 million light years. The central region of stars and dust seen in Fig. 12.10 is 20 000 light years across, but long-exposure photographs reveal fainter outer regions extending to a total size of 300 000 light years; they would cover half of Fig. 12.10. Cygnus A is the main member of a small group of galaxies. (One companion is the fuzzy blob 2 centimetres to the right of Cygnus A.) The motions of these galaxies indicate the mass of Cygnus A is a staggering 100 million million Suns – making it one of the most massive galaxies known.

The X-ray view (**Fig. 12.11**) reveals very hot gas held in by this powerful gravity. The picture covers a region about three times larger than Fig. 12.10, with the dots showing the intensity of X-ray emission. The distribution of the 100 million degree gas is also shown by the graphs, which are scans of the X-ray intensity across the

picture (the bottom graph) and up and down (the left-hand graph). The hot gas is evidently more closely packed towards the galaxy's centre, like the stars, and there is a faint extended 'atmosphere' around Cygnus A .

The most important aspect of Cygnus A is its exceptionally powerful radio emission. Whereas typical radio galaxies (like Centaurus A and M87) are a thousand times more powerful in radio waves than the Milky Way, Cygnus A is a thousand times more powerful still – equal in its radio output to a million Milky Ways. This intense emission comes not from the centre of the galaxy, but from two lobes, one on either side, as seen in **Fig. 12.12**. This image shows the radio brightness directly, and is the same width as the optical photograph (Fig. 12.10). The lobes – clouds of magnetic field and electrons – stretch out 200 000 light years on either side to terminate in bright *hot spots*.

The lobes' energy comes from the galaxy's active core – probably a disc of gases spiralling into a very massive black

hole – visible here as the tiny intense source at the centre. A pair of narrow oppositely-directed beams or jets of fast electrons carry energy outwards until they hit the surrounding gas. The orderly motions of the electrons are then disrupted; they generate magnetic fields, and as the electrons move through these fields they radiate their energy as radio waves. The collision on each side is thus marked by a hot spot in the radio map. Electrons splash back from the impacts in intricate swirls to create the large lobes behind the hot spots.

The impact of the electron beams pushes away the surrounding gas at the collision point, and so the hot spots are moving outwards, at 10 000 kilometres per second. The hot spots have hollowed out burrows which are relatively empty except for the tenuous magnetic field and electrons left behind. Cygnus A has taken five million years to grow to its current size – a short but intensely active period on the astronomical scale.

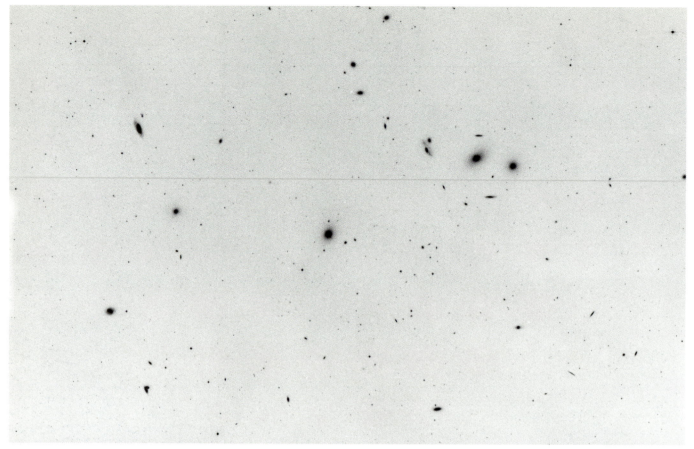

12.13 *Optical, red light, negative print, 1.2 m UK Schmidt Telescope*

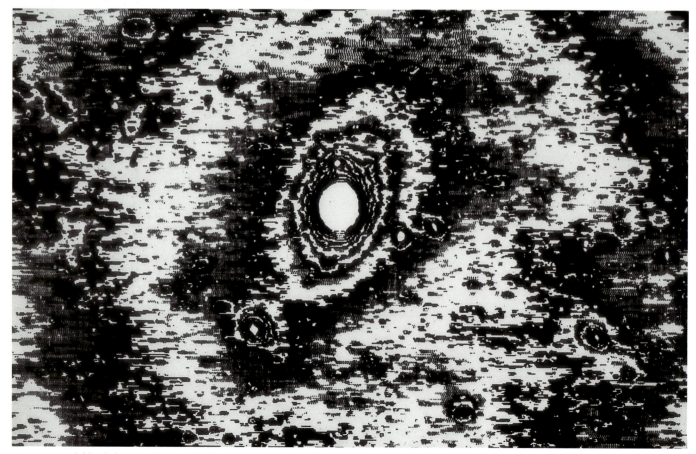

12.14 *Optical, blue light, intensity coded, 1.2 m Palomar Schmidt Telescope*

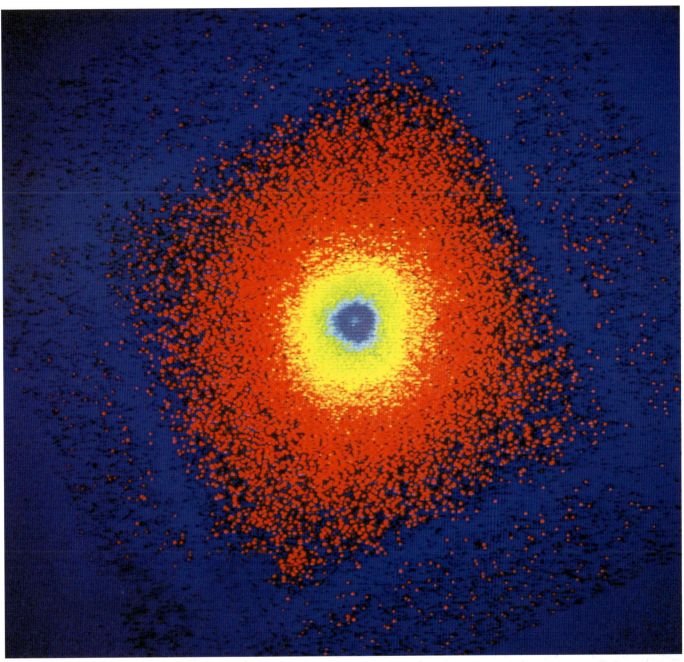

12.15 *X-ray, 0.3–2.5 nm, Imaging Proportional Counter, Einstein Observatory*

M87

SOME 50 MILLION LIGHT years away, in the constellation Virgo, lies the nearest major cluster of galaxies. The Virgo Cluster consists of a thousand galaxies concentrated into a region 10 million light years across. It is also the centre of a loose swarm of galaxy groups, the Local Supercluster, which includes our Local Group – and thus the Milky Way.

The Virgo Cluster covers 12° of sky, and **Fig. 12.13** shows just the central 5° by 3°. The six brightest galaxies are visible in even a small telescope: from left to right they are M58, M90, M89, M87, M86 and M84. The galaxy in the middle, M87, appears fairly unremarkable here, but at radio and X-ray wavelengths it outshines the other galaxies a hundred times over. It is, in fact, the most remarkable galaxy in the cluster, lying stationary at its centre while the other galaxies swarm around at speeds of up to 1500 kilometres per second.

Fig. 12.14 covers the central part of the Virgo Cluster, at three times the scale of Fig. 12.13. The brightness levels have been converted into shades of grey, so that the bands of white, grey and black are like contour lines enclosing regions of equal brightness, building up towards the most intense region at the centre (the innermost, brightest part has been left white). The faint outermost regions of M87 reach over halfway to M86 and M84, and give M87 a total size of a million light years!

In the X-ray view (**Fig. 12.15**), the fainter regions are blue, and the more intense central regions coded in red, green, purple and white. (The large blue square is caused by the detector's window supports.) The purple and white region corresponds to Fig. 12.14's central white area. The Virgo Cluster as a whole contains hot X-ray emitting gas at 100 million degrees; but the gas here – at 30 million degrees – is associated with M87 itself, and extends to half a million light years. To retain this gas, M87 must have a mass of 30 million million Suns – ten times the mass of the galaxy as deduced by the light from its stars. Most of M87's mass must be in the form of invisible dark matter.

12.16 *Optical, true colour, 3.9 m Anglo-Australian Telescope*

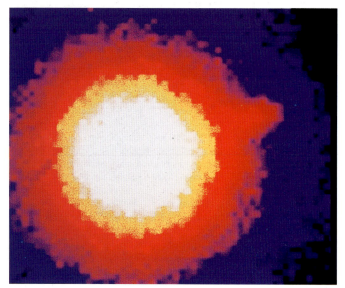

12.17 *Infrared, 2.2 μm, 3.8 m UK Infrared Telescope*

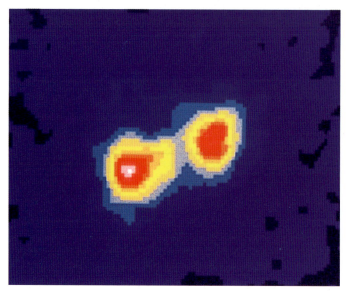

12.18 *X-ray, 0.4–8 nm, High Resolution Imager, Einstein Observatory*

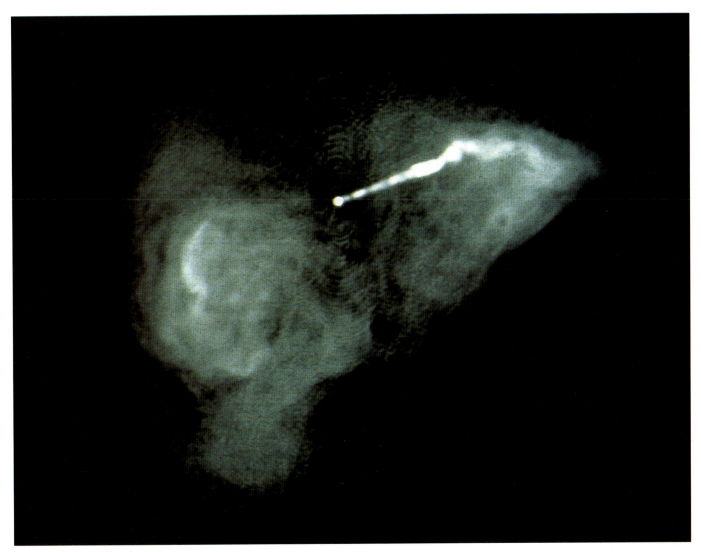

12.19 *Radio, 6 cm, Very Large Array*

The giant elliptical galaxy M87 is a stunning sight in a true colour photograph (**Fig. 12.16**), dominating its neighbouring galaxies in the Virgo Cluster. Its main body, visible here, is 50 000 light years across and packed with stars. To give an idea of its immensity, each of the small dots crowding around M87 is not an individual star but a cluster of a million stars!

In total, M87 has 4000 of these globular clusters – yet they form only a small percentage of the galaxy's total. All the millions of millions of stars making up this elliptical galaxy are old. There is very little interstellar matter in M87, and star formation has long since ceased, so there are no hot blue and white stars of the kind we find studding the arms of spiral galaxies. M87 is populated by long-lived dim red stars, and occasional red giants reaching the end of their lives.

The radiation from these cool stars shows up prominently in an infrared image. **Fig. 12.17** is coloured for infrared brightness, from white for the most intense central part down through yellow and red to blue for the dimmer regions further out.

The whole of this area lies within the extent of the galaxy as seen in the optical photograph (Fig. 12.16). Despite the elongated shape of the outer regions of M87 (Fig. 12.14), the stars in this central region form an almost spherical ball.

Superimposed on the circular outline in Fig. 12.17, however, is a distinct protuberance pointing to the upper right. This faint 'jet' was first noticed on short-exposure optical photographs by the American astronomer Heber Curtis in 1916. It excited little comment at the time, partly because no-one had the slightest idea what it was, and also because it seems quite unimportant when compared to the brightness of the massed stars in M87.

But the advent of the new astronomies in the late twentieth century changed all that. **Fig. 12.18** shows the central regions of M87 as 'seen' at X-ray wavelengths, with colour coding showing intensity increasing from purple and blue for faint emission through yellow and red to white for the brightest patch at the precise centre of the galaxy. It is to roughly the same scale as the infrared view (Fig. 12.17). The jet now

shows up as a bright spot (red) almost as brilliant as M87's core. Clearly, something remarkable is going on.

In the early days of radio astronomy, M87 was one of the first sources to be identified. **Fig. 12.19** is a recent detailed image, showing its intensity as viewed at radio wavelengths. The galaxy has a pair of radio-emitting clouds which extend to 8000 light years from the galaxy's centre, totally within M87 as seen at optical wavelengths (Fig. 12.16). It is quite a restrained size for a radio galaxy, compared to the giant radio lobes of Centaurus A (Fig. 12.4) or Cygnus A (Fig. 12.12).

Even more unusual for a radio galaxy, the twin lobes are not the most brilliant part of M87 when observed at radio wavelengths. The most prominent feature is – once again – the jet. This is the conduit through which a galaxy sends energetic electrons to the radio lobes, and in M87 many of these electrons are just not getting through. They are dumping their energy in the jet instead, making it shine at all wavelengths from radio waves right down to the shortest X-rays.

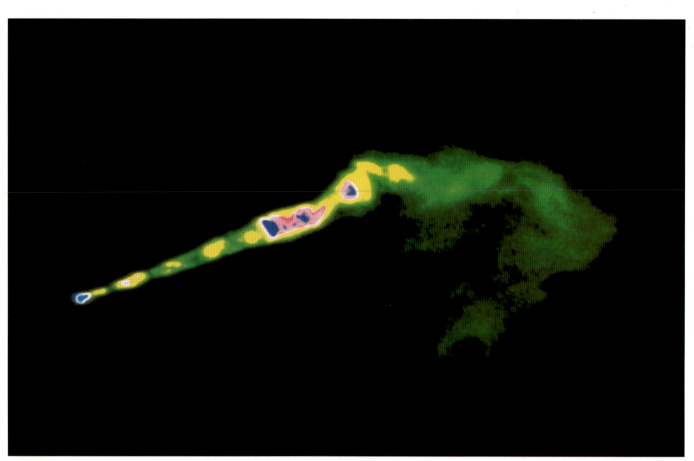

12.20 *Radio, 2 cm, Very Large Array*

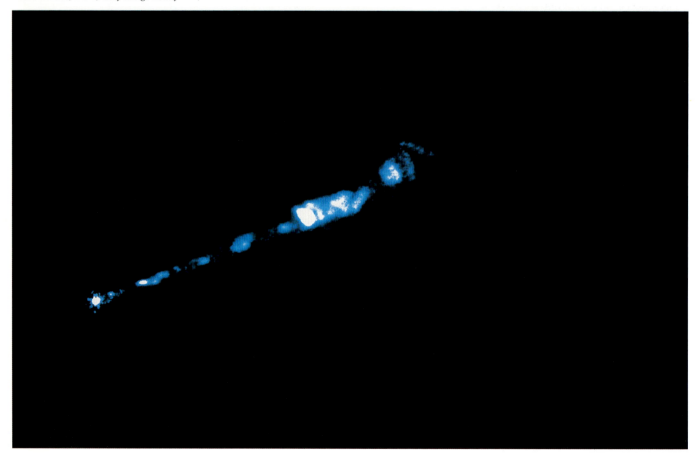

12.21 *Ultraviolet, 372 nm, Hubble Space Telescope*

220 *ACTIVE GALAXIES – AND BEYOND*

The strange radio jet of M87 is laid bare in **Fig. 12.20**. It is colour coded for intensity, with faint regions green and successively brighter regions yellow, pink, purple and blue. The tiny blue blob at the left is the galaxy's nucleus, and the jet gradually widens with distance from the centre of M87. At the right end, 8000 light years out, the jet bends around and delivers energy to the extended radio-emitting lobe on that side of the galaxy (Fig. 12.19).

The jet marks the path of high-speed electrons – possibly accompanied by protons or positrons – ejected from the centre of the galaxy. Towards the end, the jet becomes increasingly wavy. Possibly the speeding beam of matter becomes unstable here, and starts to wiggle up and down like a fire fighter's hose.

Fig. 12.20 also reveals that most of the radio emission comes from a dozen individual bright 'knots', equally spaced along the jet. These knots are moving outwards at half the speed of light. But the electrons in the jet may be moving even faster, continually flowing through and past the knots.

Support for this theory comes from studying the jet in even finer detail. The ultraviolet view of the jet (**Fig. 12.21**) is coded so that the dimmer regions of the jet appear blue and the bright knots white. Despite the enormous difference in wavelength between radio and ultraviolet, the jet looks remarkably similar. The knots appear in exactly the same places, and with similar brightness. And the ultraviolet image shows that the brightest knot actually runs *across* the line of the jet.

These observations suggest that the knots mark the positions of shock waves within the jet. They may occur where the edge of the jet is unstable: like the wiggles of the hosepipe, this kind of instability would impose a natural rhythm to the spacing between the knots. Each shock wave stops some of the electrons speeding along the jet, and converts their energy into radiation of all wavelengths.

An optical image (**Fig. 12.22**) from the sharp eye of the Hubble Space Telescope shows both the inner part of the jet and the bright glow of stars and gas in the centre of M87. It is rotated 25° anticlockwise relative to the previous images of the jet. Following the jet inwards, it points straight as a die to the galaxy's centre. And Fig. 12.22 reveals a small oval cloud of glowing gas surrounding the galaxy's very core.

This cloud appears in finer detail in **Fig. 12.23**. The gas forms a swirling disc, only 500 light years across. Its spectrum reveals the velocity of gas here. The disc turns out to be in the gravitational grip of a compact and invisible body as massive as three billion Suns. This is the best evidence to date for the theory that supermassive black holes are responsible for the awesome power of active galactic nuclei.

12.22 *Optical, 658 nm redshifted hydrogen line, Hubble Space Telescope*

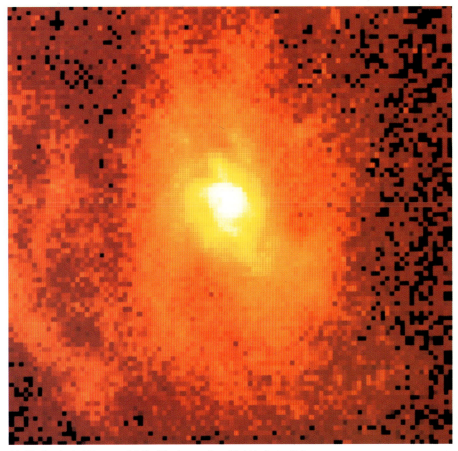

12.23 *Optical, 658 nm redshifted hydrogen line, Hubble Space Telescope*

12.24 *Optical, 3.8 m Mayall Telescope*

NGC 1275

THE RAGGED LINE of galaxies running left to right across **Fig. 12.24** contains the brightest members of the Perseus Cluster. Lying 230 million light years away, the galaxies in Fig. 12.24 span a region equal in size to our Local Group of galaxies, some five million light years across. But hundreds of fainter galaxies lie further out. Their distribution is centred near the left end of this chain of galaxies – close to the brightest galaxy, NGC 1275.

In most clusters, the central dominant galaxy is a giant elliptical. But even the small scale photograph here shows that NGC 1275 is different, with ragged wisps of gas and dust.

What makes NGC 1275 different is, almost certainly, its surroundings. Most of this region is filled with superhot gas, which shines brilliantly in the X-ray image, **Fig. 12.25**. The field of view here has a four-leafed clover shape, because the image is a mosaic of four exposures, and the supports in the detector appear as thin silhouettes. Low-brightness regions are coded blue, and successively brighter regions green, yellow and red.

The X-ray image is centred on NGC 1275, and is at only one-thirtieth the scale of the optical photograph. The central red region shows emission from NGC 1275, and the tiny yellow and red source to its right is the galaxy at the right end of the chain seen optically (IC 310).

The remaining emission (yellow and green) comes from a huge pool of gas between the galaxies, at a temperature of 80 million degrees. Held in by the cluster's gravitational pull, its slightly oval shape indicates that the distribution of mass in the cluster is mildly flattened. The total pull is ten times stronger than the galaxies can contribute, indicating a huge amount of dark matter in the cluster.

This huge halo of superhot gas is having a profound effect on the life and times of NGC 1275.

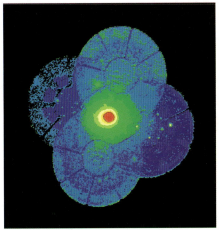

12.25 *X-ray, 0.6–10 nm, Position Sensitive Proportional Counter, Rosat*

12.26 *Optical, enhanced true colour, Wide Field/Planetary Camera, Hubble Space Telescope*

NGC 1275 drew attention to itself as far back as 1943, when an American astronomer, Carl Seyfert, compiled a list of a dozen galaxies containing a bright central point of light. All except one of these Seyfert galaxies were spirals: the exception was NGC 1275.

The tiny bright core is well seen in a view from the Hubble Space Telescope (**Fig. 12.26**). In today's terminology, it is an active galactic nucleus. If NGC 1275 were moved several times further away, we would no longer see the surrounding galaxy, and astronomers would classify its bright star-like centre as a quasar.

The Hubble image shows two other peculiarities. NGC 1275 is littered with dust; and it has around 50 bright blue star clusters (each appearing as a faint point of light here). The colour indicates these star clusters are comparatively young, dating back no more than 100 million years. How has the old galaxy NGC 1275 acquired a source of fresh gas to create this new generation of stars?

X-ray astronomy provides the surprising answer. **Fig. 12.27** is a magnified X-ray view of the region seen in red in Fig. 12.25.

The quasar core of NGC 1275 appears as the bright spot in the centre. The rest of the emission comes from dense gas at a temperature of 10 million K – much lower than the 80 million degree gas further out in the cluster of galaxies.

Astronomers explain this difference as a cooling effect. Dense gas loses energy more efficiently than tenuous gas, so the temperature of the gas near NGC 1275 has dropped. As it cools, the gas shrinks in volume and dense streamers begin to rain down on the galaxy. Gas from further out in the Perseus Cluster then moves inwards and cools in turn, to form a continuous 'cooling flow' of dense gas. Entering the inner regions of the galaxy, this gas condenses into clusters of young blue stars.

Some of the inflowing gas falls into the galaxy's very centre, where it forms a blazing disc surrounding a massive black hole: the active galactic nucleus.

Radio observations come closest to the heart of NGC 1275. **Fig. 12.28** reveals just the innermost ten light years of the galaxy, using widely scattered radio telescopes to simulate an instrument as wide as the Earth. The colour coding runs from white

for the brightest region, through red and yellow to blue for the faintest.

From the quasar core, a jet of electrons stretches downwards for several light years. It also appears in a series of observations (**Fig. 12.29**) made on five different occasions between 1972 and 1976. The brightest regions are here coded red, and successively fainter parts yellow, green and blue.

At this wavelength, the quasar core is fainter, varying in brightness as its activity changes. The jet heading downwards is clearly increasing in length from year to year, extending outwards at half the speed of light. It probably started in a major burst of activity in the nucleus, detected by radio astronomers in 1958.

As the further parts of the radio jet speed outwards, they must punch through the inflowing gas. The cavity scooped out by the jet appears in the X-ray image (Fig. 12.27) as the dark region below the galaxy's centre. Jets emitted at other periods have excavated the dark cavities above and to the upper right. So we are seeing, in a single image, both the infalling fuel for NGC 1275's activity and the aftermath.

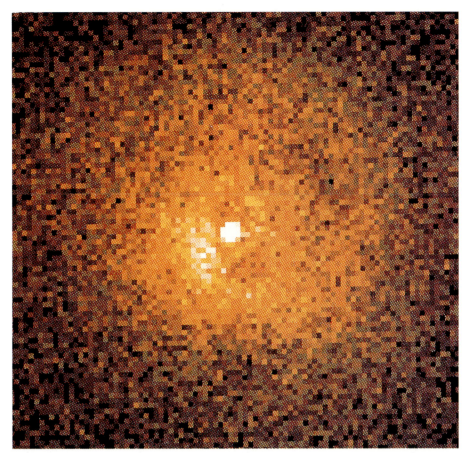

12.27 *X-ray, 0.6–10 nm, High Resolution Imager, Rosat*

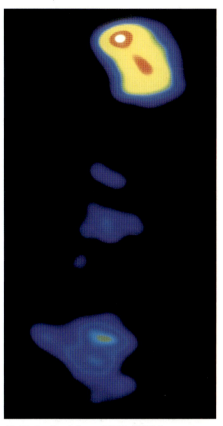

12.28 *Radio, 1.3 cm, VLBI (Effelsberg, Haystack, Green Bank, Algonquin, Very Large Array, Owens Valley)*

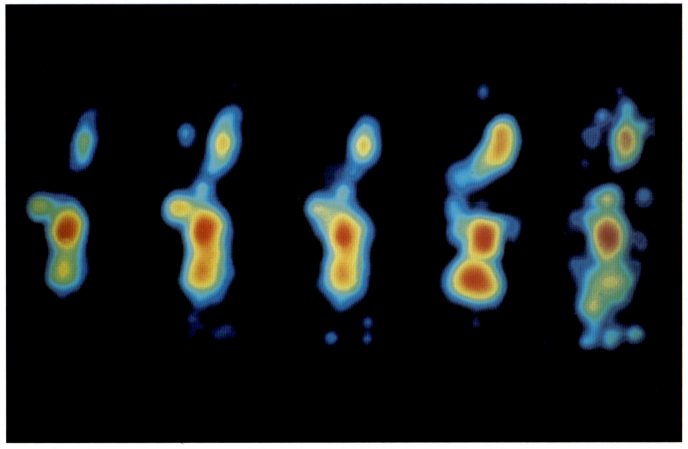

12.29 *Radio (yearly images 1972–6), 2.8 cm, VLBI (Effelsberg, Green Bank, Algonquin, Fort Davis, Owens Valley)*

3C 273

THE STAR-LIKE OBJECT at the centre of **Fig. 12.30** is the nearest quasar, apparently the brightest at all wavelengths from radio to X-ray, and the best-studied. It was first detected as the twentieth strongest radio source, and appears as number 273 in the third Cambridge catalogue. 3C 273's position was found accurately in 1962, by timing the exact moment that the Moon passed in front and blocked off the radio emission. The technique also showed that it consists of two closely spaced radio sources. On optical photographs, one radio source coincides with the bright 'star' in Fig. 12.30 and the other with the end of the faint 'jet' which extends 22 arcseconds to the lower right. The 'star' is relatively bright (at magnitude 13) and its image is overexposed here and blurred out to an artificially large size (as are other images of true stars).

Astronomers thought at first that 3C 273 was indeed a star in our Galaxy, but its spectrum turned out to be very different. Dutch-American astronomer Maarten Schmidt realised in 1963 that the lines seen crossing the spectrum of 3C 273 are due to hydrogen – the commonest element – but all with their wavelengths increased substantially. The strongest spectral line from hydrogen occurs at ultraviolet wavelengths, and this line (Lyman-alpha) was studied in detail by the International Ultraviolet Explorer satellite in 1978.

Fig. 12.31 shows a small part of 3C 273's ultraviolet spectrum running from the top left to the lower right. (The jet is too faint to show up in the spectrum.) The superimposed red dotted line marks the centre of the spectrum, with wavelengths marked in angstrom units from 1200 (120 nanometres) to 1400 (140 nanometres), and intensities coded from dark blue for the background, through pale blue to white for the brightest regions. The intense spectral line next to the figure 1200 is hydrogen's Lyman-alpha line at its normal wavelength (122 nanometres), and this comes from hydrogen surrounding the Earth in a huge, hot cloud called the geocorona. The Lyman-alpha from 3C 273 appears in the lower right-hand corner near the figure 1400, with its wavelength increased by 19 nanometres. The proportional increase is 16 per cent, and the same value is found for all of 3C 273's spectral lines.

The simplest explanation for the shift in the spectral lines is that 3C 273 is so far away from us that the expansion of the Universe is carrying it away at a speed of 50 000 kilometres a second, and the Doppler effect thus stretches all the wavelengths. 3C 273 must then lie at a staggering 2100 million light years away and, to appear as a relatively bright 'star' in our skies, it must be more luminous than the brightest galaxies known, as brilliant as

12.30 *Optical, 3.8 m Mayall Telescope*

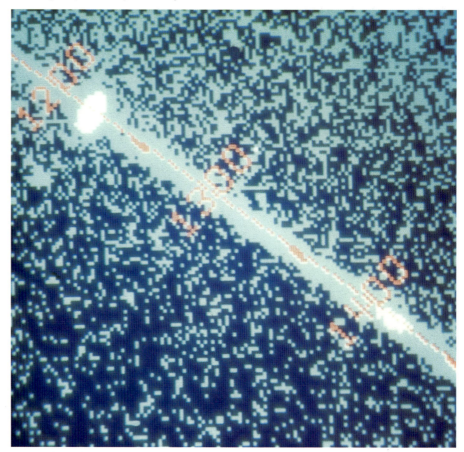

12.31 *Ultraviolet, spectrum, 120–140 nm, International Ultraviolet Explorer*

400 Milky Ways. In addition, the light comes from a very small region. 3C 273's star-like appearance on photographs, and indirect arguments based on changes in its brightness, indicate that it is less than one-hundredth of a light year across.

Since 3C 273 was clearly not a star, despite its appearance, it was dubbed a 'quasi-stellar radio source', later abbreviated to 'quasar'. Thousands of other, more distant quasars, are now known. Further investigations have shown that quasars are outbursts in the centres of distant galaxies, similar to the smaller outbursts found in galaxies like NGC 1275 (Fig. 12.26).

Quasars are so brilliant that it is usually difficult to detect the surrounding galaxy. In **Fig. 12.32**, 3C 273 has been recorded by a very sensitive CCD light detector, with a telescope on the high summit of Mauna Kea, Hawaii, where images of astronomical objects are least blurred by Earth's atmosphere. Levels of brightness are colour coded, from black for dark sky, through blue, red and yellow to black for the brightest regions.

The four 'spikes' apparently emerging from 3C 273 are caused by struts in the telescope. The jet is now clearly revealed, extending to the lower right and detached from the quasar. Most interesting, the brilliant quasar (coded black) is clearly surrounded by an oval-shaped galaxy. Although only one-twentieth as bright as its quasar core, this ranks with the largest and brightest elliptical galaxies known: it outshines the Milky Way twenty times, and is three times as large, with a diameter of one-third of a million light years – similar to Cygnus A (Fig. 12.10).

3C 273 is also a strong X-ray emitter, producing as much power in X-rays as it does in light, and is over a million times more powerful at this wavelength than the Milky Way. Early X-ray observations showed that its output can vary in half a day, indicating that this huge amount of radiation comes from a region no larger than the Solar System. The Einstein Observatory looked at 3C 273 in roughly the same detail as early optical photographs like Fig. 12.30. In **Fig. 12.33**, the X-ray brightness is colour coded, with dark blue, red and brown corresponding to the darkest regions, and successively brighter regions in buff, purple, grey, yellow, green and pale blue. In this long-exposure picture, 3C 273 is smeared out to an artificially large size, and the background has false 'spokes' caused by the supports of the telescope mirrors. But the yellow and green blob at 'four o'clock' is real, and corresponds to the inner half of the jet as seen at optical wavelengths. This beam of electrons emits only one three-hundredth as much light and X-rays as the quasar's core. But even so, the jet alone emits as much light as our Milky Way Galaxy, and is 10 000 times more powerful in X-rays.

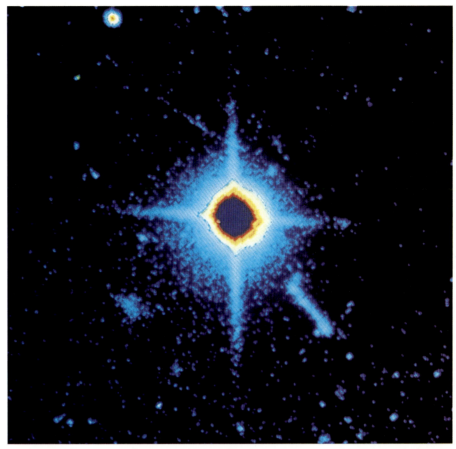

12.32 *Optical, 3.6 m Canada–France–Hawaii Telescope*

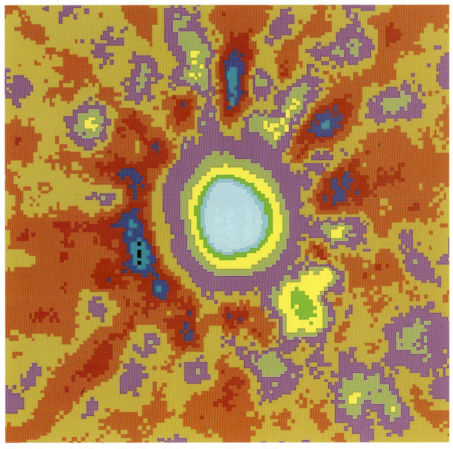

12.33 *X-ray, 0.4–8 nm, High Resolution Imager, Einstein Observatory*

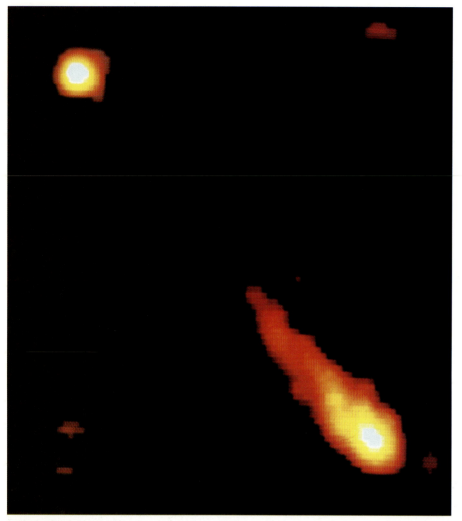

12.34 *Radio, 73 cm, MERLIN*

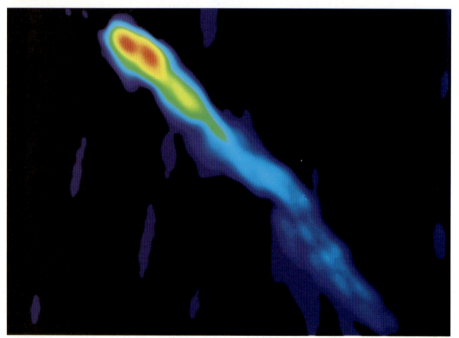

12.35 *Radio (March 1986), 18 cm, World VLBI Network (Lovell Telescope, Defford, Cambridge (UK); Onsala (Sweden); Medicina (Italy); Effelsberg (Germany); Westerbork (Holland); Crimea (Russia); Green Bank, North Liberty, Pie Town, VLA, Owens Valley, Haystack (USA); Arecibo (Puerto Rico); Hartebeesthoek (South Africa)*

As seen at radio wavelengths, the quasar 3C 273 has a totally different look. **Fig. 12.34** is colour coded so that bright regions are white, shading into yellow, orange and red for fainter parts (the small isolated red patches are false images). The elongated jet at lower right is every bit as powerful as the quasar itself (upper left), each emitting a million times the Milky Way's radio output. The jet seen at radio wavelengths coincides in position and length with the optical jet, but the radio jet is a hundred times more powerful at the outer than the inner end.

In fact, an in-depth comparison of the optical (Fig. 12.32), X-ray (Fig. 12.33) and radio (Fig. 12.34) observations reveals how and why the jet is emitting its energy. Like the jets seen in other galaxies – and some stars in our Galaxy – it undoubtedly consists of a narrow stream of electrons, beamed out from the galaxy's brilliant quasar nucleus. Once the electrons leave the region of the nucleus, they coast along for over 100 000 light years without emitting any energy. This inner region of the jet is invisible at any wavelength.

Then the jet hits a region of gas in the outer part of the giant 3C 273 host galaxy – possibly the remains of a small galaxy that it has swallowed up. The shock knocks some of the electrons out of the overall outward flow. Their random motion generates a tangled magnetic field, and the electrons' motion within the lines of magnetism generates radiation by the synchrotron process.

At the location of the shock, this emission covers all wavelengths, and is particularly strong in the short optical and X-ray region. But the high-energy electrons that produce X-rays lose their energy very rapidly. The emission at short wavelengths thus occurs mainly near the shock. The electrons shining at visible wavelengths survive for longer and so can propagate further along to create the long jet seen in optical images.

As all these electrons lose energy, they end up emitting long-wavelength radio waves. These low-energy electrons can continue broadcasting for almost a million years, and they end up at the far end of the jet. During this long lifetime, they can also diffuse sideways, so the radio jet is wider.

The result of these processes is that the inner part of the jet is bright at X-ray wavelengths; the optical jet is narrow and fairly uniform all along; while the radio jet is brightest (and widest) at the end furthest from the galaxy.

But what of the central 'engine' that is producing the jet in the first place? In recent years, radio astronomers have been able to probe this core – seen blurred out at the top left of Fig. 12.34 – in incredible detail, by combining radio telescopes around the world in the technique of Very Long Baseline Interferometry (VLBI).

Fig. 12.35 is a radio image of the nucleus of 3C 273, magnified 300 times relative to the previous radio view (Fig. 12.34). Here the colour coding shows fainter regions in blue, and brighter parts in green, yellow and red. The actual centre of 3C 273 is the red blob at the extreme upper left. Even these detailed observations blur out its actual very tiny size: they show that the active core is less than 20 light years across.

From this core, the quasar is emitting a jet that is bright at first in radio waves, but gradually fades as it moves away from the nucleus. At first, the electrons have some random motion, and generate magnetic fields and synchrotron emission. As the electrons head outwards, however, their motion settles down and the jet fades from sight – as seen here – some 700 light years out. It will travel for 100 times this distance as an invisible beam of energy before lighting up again.

For a beam of energy that is settling down to a uniform flow, the tiny central jet seen in Fig. 12.35 looks surprisingly wavy. This is probably just a matter of geometry: astronomers believe the electron beam is coming almost directly towards us. In this orientation, any tiny irregularities look much larger. A carpenter uses exactly the same geometrical trick when looking end on at a wooden edge to test its straightness.

Another set of radio VLBI images (**Fig. 12.36**) shows the very core of 3C 273 at higher magnification still – and reveals action taking place even as we watch. These four images were made over a period of two years. In each, the dimmest parts are coded purple, and brighter regions blue, green, yellow and red. The oval shapes are an artefact of the observing technique. The quasar core is the red blob (upper left).

In this series of images, a radio-emitting blob is evidently moving away from the nucleus, down the line of the jet. In the first image, the blob is 4 light years from the nucleus. That is only the distance of the Sun from the nearest star, seen at a distance of 2100 million light years! By the final image, it has moved out by another ten light years. But that has taken place in less than two years: a quick sum demonstrates that this blob is apparently moving outwards at six times the speed of light.

According to Einstein's theory of relativity, nothing can move faster than the speed of light. But the 'superluminal' velocity in 3C 273 can – fortunately – be explained away by a geometrical effect. This requires the electrons to be moving just slower than light, and the jet to be orientated almost directly towards us.

If the outermost part of the jet – hundreds of times further from the nucleus – follows the same orientation, then it must also be foreshortened. Already impressive enough in scale, this would mean the jet's total length could be over a million light years.

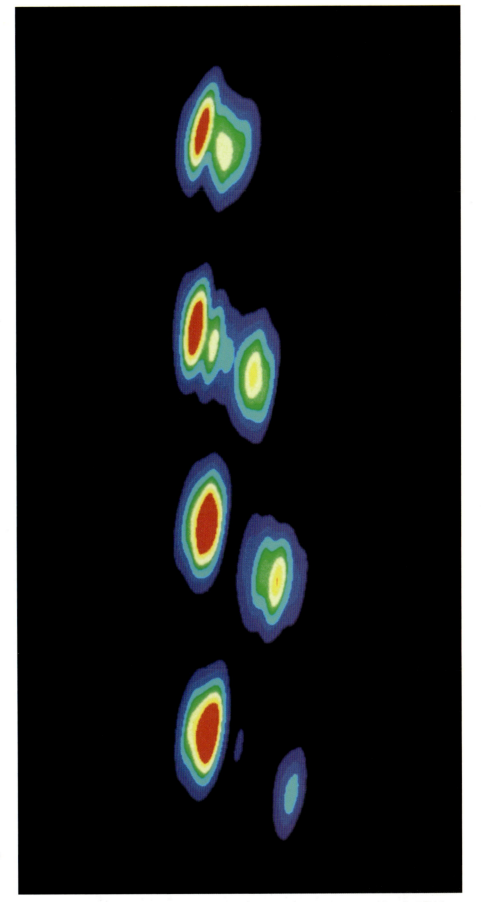

12.36 Radio (October 1985, May 1986, November 1986, May 1987), 1.3 cm, VLBI (Onsala, Effelsberg, Medicina, Crimea, Haystack, Algonquin, Maryland Point, Green Bank, VLA, Owens Valley)

Gravitational lenses

FAR OUT IN the depths of the Universe, things are not always what they seem. In the past few years, astronomers have tracked down cosmic mirages of many kinds. Some appear as curved arcs or complete rings, others as multiple images of a single quasar or galaxy. The effect is sometimes best shown up by a radio telescope, and sometimes seen best with an optical telescope. A comparison of the illusion at different wavelengths helps to pin down its cause.

The ultimate culprit is always gravity. Albert Einstein's general theory of relativity, published in 1916, predicted that a gravitational field can bend the path of light. Astronomers confirmed the theory when they found that the Sun's gravity deflected passing starlight by a tiny amount during a total solar eclipse in 1919.

The depths of the Universe provide a vast stage for testing two other predictions of Einstein's theory. First, radiation of all wavelengths should be bent equally by gravity. And secondly, a massive object should not just bend light (or other radiation) from an object behind, but focus it to form a distorted image. Depending on the details of this 'gravitational lens', it may produce multiple images or curved arcs.

Appropriately, it was during the centenary of Einstein's birth that astronomers stumbled across the first gravitational lens. Radio astronomers at Jodrell Bank had compiled a list of distant radio sources in the 1970s, and in 1979 they checked out the optical appearance of one source that – in their radio survey – seemed no different from the others. They suspected it would be a distant quasar.

Fig. 12.37 was the sight that met their astonished eyes. They seemed to be seeing double. Instead of one quasar, there were two, almost equal in brightness and just a few arcseconds apart. This colour-coded image mimics a photographic negative, with the background sky blue, brighter regions green, yellow and red, and the overexposed brilliant centres of the quasars black. (The quasar images look elongated in this highly magnified view because the telescope mirrors were very slightly out of line.)

The 'twin quasar' became curiouser and curiouser when the team studied the spectra of the two objects. The light from quasars is generally very diverse, and when split up into a spectrum each has a unique set of bright emission lines, differing in strength and width from every other quasar. The lines are also redshifted to longer wavelengths by an amount that depends on the quasar's distance from us in the expanding Universe. But the spectra of the two quasars in Fig. 12.37 turned out to be absolutely identical. They must be two

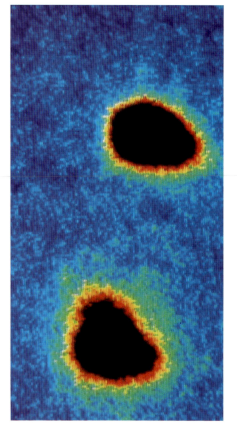

12.37 *Optical, red light, intensity coded, 2.2 m reflector, Mauna Kea*

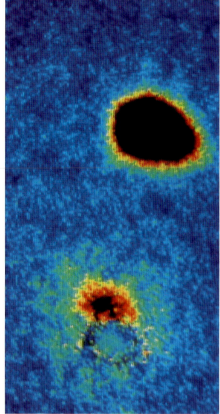

12.38 *Optical, red light, intensity coded with lower quasar image subtracted, 2.2 m reflector, Mauna Kea*

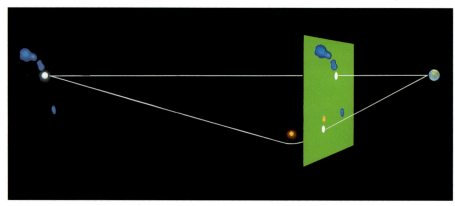

12.39 *Diagram, gravitational lensing of the twin quasar (angular scale exaggerated)*

images of the same quasar.

The shift in the spectral lines indicates that this quasar lies some 10 000 million light years away. It is a fairly average specimen, one-tenth as luminous as 3C 273 (Fig 12.30) and fifty times brighter than the Milky Way. Its brilliant emission probably comes from a hot gas disc around a massive black hole, in a galaxy too faint to be discerned at this distance.

According to Einstein's theory, the quasar's light can only be focused if it passes close by a galaxy lying right in front, as seen from the Earth. A galaxy would be difficult to detect in front of the brilliant – though more distant – quasar. But there is a hint: the lower quasar image in Fig. 12.37

appears slightly fuzzy at the top left edge.

Since the two quasar images should be identical, we can use a computer to subtract away a copy of the top image (quasar alone) from the bottom image. The result is shown in **Fig. 12.38**. The subtraction has indeed removed the lower image of the quasar entirely, and we see a fuzzy blob: the galaxy responsible for focusing the quasar light.

The galaxy lies 4000 million light years from us, about one third of the way to the quasar. It is a giant elliptical galaxy, of the same type as M87 (Fig. 12.16), some ten times brighter than the Milky Way and a hundred times more massive. The black and red regions in Fig. 12.38 show its inner

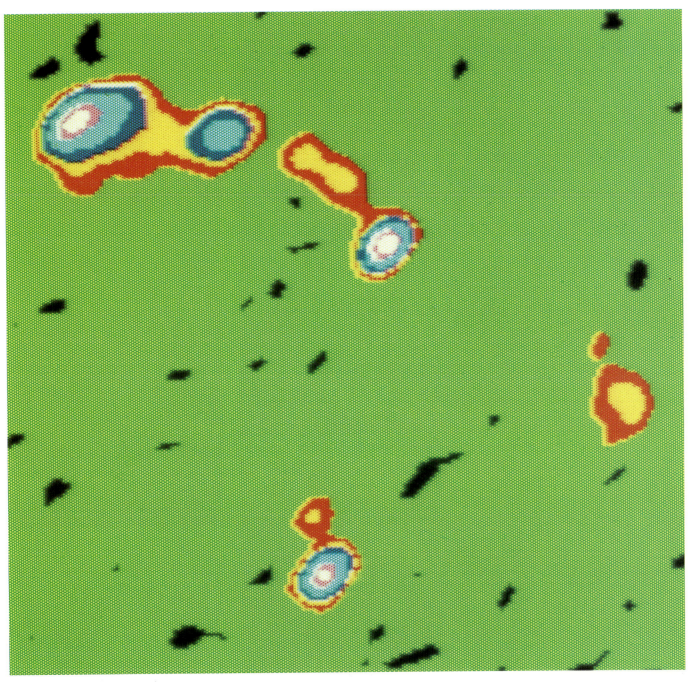

12.40 *Radio, 6 cm, Very Large Array*

20 000 light years, but the surrounding green halo reveals it extending to at least 100 000 light years. It is the central, brightest member of a cluster of at least 100 galaxies. The combined gravitational pull of the cluster adds to the central galaxy's lensing effect.

The first radio observations revealed only the position of this object, but a highly detailed radio view (**Fig. 12.40**) shows a rather complex structure. Here, the background sky is coded green and black, with regions of successively more intense radio emission coded red, yellow, dark blue, light blue, purple and white.

The small white-centred source above the middle coincides with the upper image

seen optically: it is the quasar's tiny core. To either side of this source, Fig. 12.40 shows a lobe of radio emission. The lobes stretch out 100 000 light years in both directions, similar in size to the lobes of Cygnus A (Fig. 12.12).

The white-centred source at the bottom of Fig. 12.40 is at exactly the same position as the lower image of the quasar in the optical photograph. And, like the optical images, the two are roughly equal in brightness. Just as Einstein predicted, the gravitational lens is treating the radio waves in exactly the same way as light, even though the wavelength is 100 000 times longer.

Fig. 12.39 shows the geometry. Earth is

at the right (with our view depicted on the screen) and the quasar (white spot) at the left with its two radio lobes (blue). The intervening galaxy (yellow) has little effect on the radiation travelling directly to Earth (top of screen), but it bends light and radio waves travelling near the galaxy's centre so much that it creates a second image of the quasar's core.

The radio view (Fig. 12.40) reveals a weak source (red and yellow) above the lower quasar image, coinciding with the centre of the lensing galaxy. This could be radio emission from the core of the galaxy itself, or it could be another mirage: a fainter third image of the quasar behind.

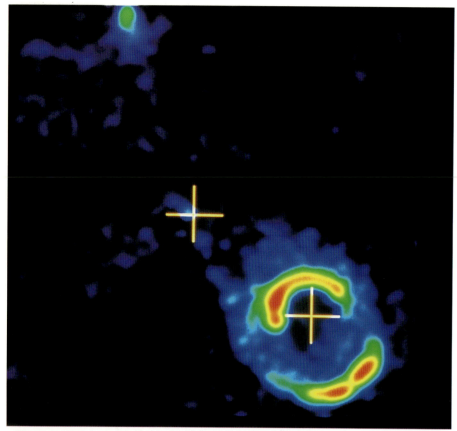

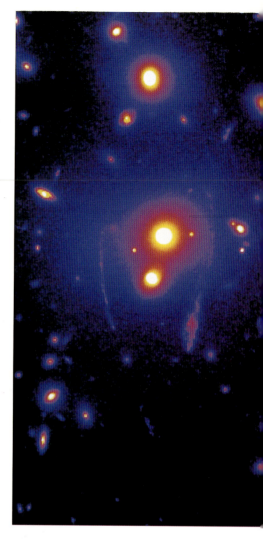

12.41 *Einstein ring MG 1654+1346. Radio, 3.6 cm, Very Large Array*

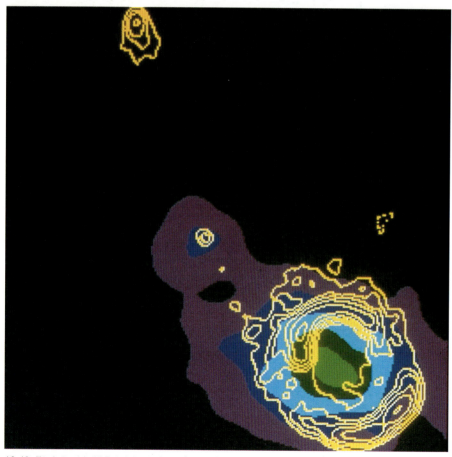

12.42 *Einstein ring MG 1654+1346. Radio contours on optical red light image, intensity coded, 1.3 m reflector, McGraw-Hill Telescope, Kitt Peak*

Radio observations put astronomers on the trail of one particularly fine gravitational mirage. In 1988, they turned the Very Large Array to an obscure radio source known only by its coordinates, MG 1654 + 1346. It turned out to have a unique radio structure (**Fig. 12.41**). Here, the colour indicates the intensity of radio emission, with fainter parts blue and brighter regions green, yellow and red. Most of the radio emission is coming from a small, almost circular ring. There is also a much fainter and smaller radio lobe at the top left, and a compact source near the centre. (The crosses mark the positions of two objects seen at optical wavelengths.)

It takes a multi-wavelength comparison to reveal what is going on. In **Fig. 12.42**, the radio brightness from Fig. 12.41 has been converted to contour lines. The coloured part of the image is the optical view, coded for brightness. Purple represents the faintest light levels, with brighter regions dark blue, pale blue, dark green and light green. There is clearly a bright extended object at the lower right, in the middle of the ring of radio emission, and a much fainter object (blue patch) in

12.43 *Gravitational lens Abell 2218. Optical, intensity coded, Wide Field/Planetary Camera 2, Hubble Space Telescope*

the centre of the field of view, coinciding with the weak radio source.

By examining these objects with a spectrograph, astronomers have found that the large object is a galaxy at a distance of 1700 million light years, while the fainter one is a quasar far in the background.

Now the truth becomes apparent. The bright radio source is merely a mirage: an almost perfect example of an 'Einstein ring.' According to Einstein, if a massive body lies exactly in front of a source of radiation, its gravity will focus the radiation into a much brighter circular halo.

Here, the distant quasar actually has two rather weak radio lobes: the region we see at the top left of Fig. 12.41 and a similar lobe to the bottom right. But, purely by chance, the galaxy happens to lie right in front of the second lobe. Its gravity focuses the radio waves from this lobe into a bright Einstein ring.

The size of the Einstein ring reveals the strength of the gravitational lens. This is the only direct method of weighing up such a distant galaxy. Its mass turns out to be 300 000 million Suns. If all this matter were in the form of stars, the galaxy would

shine much more brightly than we see it. So a substantial amount of this mass must be in the form of invisible 'dark matter'. It could be in the form of dark lightweight stars, or possibly black holes. But, most likely, it consists mainly of a massive 'sea' of subatomic particles.

The powerful gravity of a whole cluster of galaxies can act as a huge gravitational lens. If we look through a nearby cluster of galaxies, we may see a distorted view of anything that lies beyond.

A magnificent example was captured by the Hubble Space Telescope in 1994. **Fig. 12.43** (with North to the right) is colour coded for intensity: faint regions are blue, and brighter parts pink, red, yellow and white. The lens comprises the spiral and elliptical members of the comparatively nearby cluster Abell 2218, which appear as relatively large and symmetrical galaxies here (along with the invisible dark matter in the cluster).

The spider's web of curved lines around the cluster is an illusion. These curved arcs are distant galaxies, five to ten times further away than Abell 2218. Because these galaxies do not lie centrally behind

the cluster, their light is not focused into a complete circle, but forms only part of an Einstein ring. Even so, the lensed galaxies are stretched into totally unnatural shapes.

The original image contains 120 different mirages. The Einstein lens splits seven of the background galaxies into two different arcs, on opposite sides of Abell 2218. This plethora of images can reveal in some detail how mass is spread through the cluster, and hence indicate the distribution of dark matter.

Gravity acts as a lens in more senses than one. As well as distorting the images in Fig. 12.43, it makes them much brighter: without Abell 2218, these background galaxies would not be visible at all. The combination of the giant natural lens and Hubble's superior view is showing galaxies 50 times fainter than any ground-based telescope could hope to achieve.

We see the distant lensed galaxies here as they were when the Universe was one-quarter its present age. The internal structures in the images – once they have been corrected for the distortions – can provide new insights into the youth of galaxies like our own Milky Way.

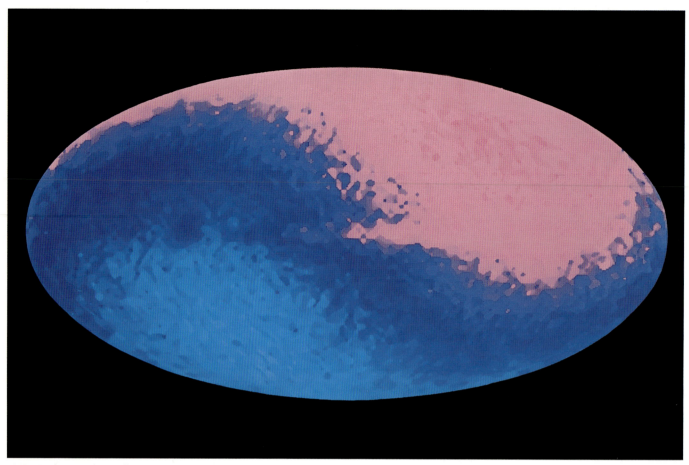

12.44 *Radio, 5.7 mm (Milky Way emission subtracted), Cosmic Background Explorer*

The microwave background radiation

THE NEW ASTRONOMIES of the late twentieth century have taken astronomers to the point where they can literally see the beginning of the Universe.

The key to this feat of time travel is the finite velocity of light, a speed limit shared by all other kinds of electromagnetic radiation. If we observe a galaxy ten million light years away, for example, we see it as it was ten million years in the past. The Big Bang happened some 15 billion years ago, so if we look out 15 billion light years we can expect to see the Big Bang.

Radio telescopes have enough sensitivity to detect the faintest signals from such far distances. Back in 1965, two American astronomers, Arno Penzias and Robert Wilson, unwittingly stumbled across a background 'hiss' of radio waves that came from all directions in the sky. Other astronomers quickly confirmed the discovery. They found the radiation is stronger around a wavelength of 1 millimetre, in the microwave part of the radio spectrum.

This 'microwave background radiation' is the echo of the Big Bang, resounding round the Universe to this day. It began as a blaze of visible light, in the great fireball erupting from the Big Bang. But the expansion of the Universe has stretched its wavelengths from visible, through infrared and on to the beginning of the radio spectrum. The radiation, which started at thousands of degrees, has cooled to the point where it bathes everything in a universal bath at a temperature of just 2.735 K – less than three degrees above absolute zero.

Because all the matter in the Universe erupted simultaneously in the Big Bang, the background radiation is the same in every direction we look. In fact, the very uniformity of the background radiation made the first few years of measurements a fairly dull affair. The main excitement came from the experimenters themselves: the background radiation is strongest at wavelengths absorbed by the atmosphere, so the researchers had to fly miniature radio telescopes to high altitudes, using balloons or – in one case – a converted U2 spyplane.

But more precise measurements started showing an interesting effect: one half of the sky is warmer than the other half. This difference is seen dramatically in **Fig. 12.44**. It is a recent measurement from the Cosmic Background Explorer (COBE) satellite, launched in 1989 to observe the microwave background with much higher precision than ever before.

Fig. 12.44 shows the whole sky, projected onto an oval frame with the Milky Way running horizontally across the centre, like the optical image of the Milky Way in Fig. 8.1. The colour coding shows slight differences in temperature from the average, pink for hotter regions and blue for cooler regions.

The COBE researchers have subtracted away as best they can the emission from our Galaxy (though the pink 'spike' near the centre shows where this subtraction has not worked completely). If we subtracted the Milky Way from an *optical* view of the sky, we would be left with an almost completely black background. But in the microwave region (Fig. 12.44) the radiation from the Big Bang keeps the whole Universe aglow.

The differences in temperature in Fig. 12.44 are minute by any standard. The warmest regions (dark pink) are just 0.007 degrees hotter than average and the coolest parts (pale blue) colder by a similar amount. That represents a total range of less than one per cent.

But the microwave background is actually even more uniform than this. The pattern of warm and cool hemispheres in Fig. 12.44 is not intrinsic to the background radiation, but is entirely due to our motion. Radiation that we meet head-on appears to have a slightly higher intensity, and thus an increased temperature. Radiation catching up with us

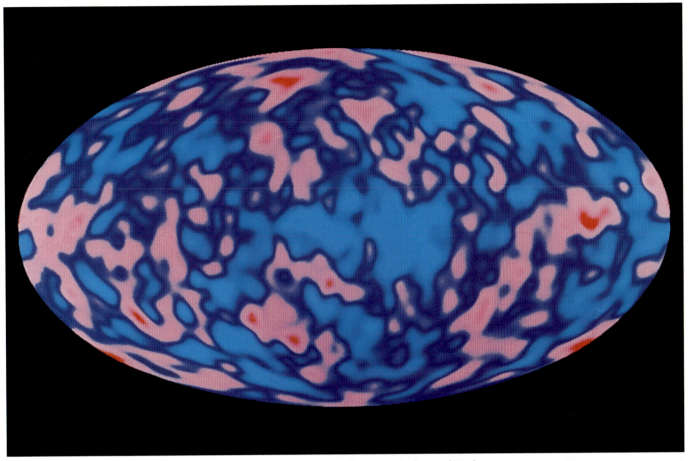

12.45 *Radio, 5.7 mm (Milky Way emission and dipole component subtracted), Cosmic Background Explorer*

from behind is dimmed and appears cooler.

The hottest region of sky in Fig. 12.44 lies in the constellation Leo. The apparent increase in temperature here means the Sun is speeding through space in this direction at 360 kilometres per second. When we allow for the Sun's orbit around our Galaxy, and the motion of the Milky Way relative to its neighbour galaxies, it turns out that the entire Local Group of galaxies is heading towards the constellation Hydra (a neighbour of Leo) at 600 kilometres per second.

What is pulling the Milky Way and its neighbours through space at such a lick? The huge Virgo Cluster of galaxies (Fig. 12.13) undoubtedly shoulders much of the responsibility. But we are not moving precisely in the direction of Virgo, and astronomers are searching for signs of another Great Attractor – a more massive and presumably more distant concentration of mass.

Once the COBE researchers had subtracted away the large-scale pattern due to our motion, the microwave background again become a boringly uniform glow. The sky was at exactly the same temperature all over, with no patches deviating by even 0.01 per cent. But a year of continuous observing enabled them to measure the temperature with ever increasing precision. And, in 1992, they

found the 'Holy Grail' of cosmology.

The COBE results, **Fig. 12.45**, hit the world's headlines as 'the ripples from the Big Bang'. The image once again shows deviations from the average temperature across the whole sky, in the same projection as Fig. 12.44. But now the colours show far smaller temperature differences. Red regions are just 30 millionths of a degree warmer than average, and the palest blue areas cooler by the same amount.

Fig. 12.45 is revealing the structure of the fireball in which the Universe was born. The patchwork of temperature shows gas from the Big Bang starting to break up into distinct lumps. As time goes by, the denser regions will condense into clusters of galaxies.

But there is a problem. Optical astronomers have found distant galaxies and quasars that seem fully formed just a billion years after the Big Bang. The modest fluctuations in Fig. 12.45 do not have enough gravity to condense within this comparatively short time.

There is only one easy way out. It invokes the idea that most of the mass in the Universe exists as 'dark matter' rather than ordinary electrons, nuclei and atoms. The dark matter curdled into clumps immediately after the Big Bang. These clumps exerted a strong pull on the hot

gases in the fireball erupting from the Big Bang, but the intense radiation in the fireball kept the gas fairly evenly spread out. Even the most concentrated regions of gas were only 10 parts in a million denser than the average for the fireball.

After 300 000 years, the fireball cooled and its gases became transparent. This is the phase of the explosion that COBE is witnessing. The small differences in density show up as correspondingly tiny variations in temperature. But after this point, radiation spread freely through space. Unsupported by radiation, the hot gas fell into the 'gravitational wells' provided by the dark matter, and condensed into galaxies, stars and planets.

Even the epoch-making COBE results do not tell us what the dark matter is made of. Some may be in the form of small dark stars or black holes, but most is likely to consist of subatomic particles – maybe neutrinos, or more exotic creatures with names like axions and gravitinos.

But the view in the final image of this book sums up perhaps the most important impact of the new astronomy. With innovative telescopes, aided by access to space, astronomers have reached out not just to explore the whole of space, but to reach back in time and examine the entire history of the Universe.

picture credits

Bold figure numbers indicate that the picture is available from the Science Photo Library, 112 Westbourne Grove, London W2 5RU. Telephone +44 (0)171 727 4712. Fax: +44 (0)171 727 6041.

1.1 NRAO/AUI. Observers: P. Scheuer, R. Laing, R. Perley
1.2 Julian Baum
1.3 NASA
1.4 Original plate taken in Egypt, 1910, supplied by D. Klinglesmith III, NASA/Goddard Space Flight Center
1.5, 1.6, 1.7 Image-processing by D. Klinglesmith III & J. Rahe, using the laboratory for Astronomy & Solar Physics' Interactive Data Analysis Facility at NASA/Goddard Space Flight Center
1.8 Martin N. England
1.9, 1.10 Julian Baum

2.1 NASA
2.2 Mount Wilson Observatory photograph, © California Institute of Technology
2.3 NRAO/AUI. Observers: G.A. Dulk & D.E. Gary
2.4, 2.5 NASA
2.6 Mount Wilson Observatory photograph. © California Institute of Technology
2.7 R. Stachnik, P. Nisenson & R. Noyes, Harvard-Smithsonian Center for Astrophysics, 'Speckle Image Reconstruction of Solar Features', Astrophys. J. Lett., 1983
2.8 National Solar Observatory, NOAO
2.9 D. Gezari, NASA Goddard Space Flight Center
2.10 NASA
2.11 C. Lindsey, Solar Physics Research Corporation, & R. Harrison, Rutherford-Appleton Laboratories
2.12 L. Golub & G.S. Vaiana, Harvard-Smithsonian Center for Astrophysics
2.13 J. Dürst, Swiss Federal Observatory
2.14 Japan Institute for Space & Astronautical Science (JISAS) & NASA
2.15 SAO/HAO, courtesy of L. Golub
2.16 NASA
2.17 © Anglo-Australian Observatory, photograph by David Allen
2.18 NASA
2.19 D.P. Anderson, Southern Methodist University/NASA
2.20, 2.21 NASA
2.22 J. Tennyson & S. Miller, University College London
2.23 A. Dollfus, Observatoire de Paris, Meudon, 'Une Nouvelle Méthode d'analyse polarimétrique des surfaces planétaires', C. R. Acad. Sci. Paris, 1990
2.24 J. Tennyson & S. Miller, University College London
2.25 A. Metzger, D. Gilman, K. Hurley, J. Luthey, H. Schnopper, F. Seward & J. Sullivan, 'Detection of X-rays from Jupiter', J. Geophys. Res.
2.26 J.A. Roberts, University of California, Berkeley, G.L. Berge, California Institute of Technology & C. Bignell, US National Radio Astronomy Observatory; NRAO/AUI
2.27 M. Mendillo, Boston University; M. Mendillo, J. Baumgardner, B. Flynn & W.J. Hughes, 'The extended sodium nebula of Jupiter', Nature 348, 22 November 1990
2.28, 2.29, 2.30 NASA
2.31 Space Telescope Science Institute & NASA/ESA
2.32 © 1987, Royal Observatory Edinburgh
2.33 F. Espenak, NASA Goddard Space Flight Center
2.34 NRAO/AUI. Observer: I. de Pater
2.35 Barney Magrath
2.36 Naval Research Laboratory
2.37, 2.38 European Space Agency
2.39 NRAO/AUI. Observers: I. de Pater, P.E. Palmer & L.E. Snyder

2.40 Michael F. A'Hearn, University of Maryland; image processing by Daniel Klinglesmith, NASA Goddard Space Flight Center
2.41 C. M. Telesco & R. Drecher, NASA Marshall Space Flight Center, H. Campins, Planetary Science Institute, & D.P. Cruikshank, University of Hawaii; image processing by C. Benson & C. Sisk, NASA/MSFC

3.1 Julian Baum
3.2 © 1989 Royal Observatory, Edinburgh. Photo by B.W. Hadley
3.3 David Parker
3.4, 3.5 © Roger Ressmeyer, Corbis
3.6 © Royal Observatory, Edinburgh. Photo by B.W Hadley
3.7 John Walsh
3.8 © Anglo-Australian Observatory
3.9 © Anglo-Australian Observatory, photograph by David Malin
3.10 © Anglo-Australian Observatory
3.11, 3.12 Jean Lorre, NASA/Jet Propulsion Laboratory
3.13 David Parker
3.14 NASA
3.15 Kirk Borne, Space Telescope Science Institute
3.16, 3.17, 3.18 © Roger Ressmeyer, Corbis

4.1 Ian Gatley, National Optical Astronomy Observatories
4.2 Harvard College Observatory
4.3 Max-Planck-Institut für Extraterrestrische Physik
4.4 © Anglo-Australian Observatory, photograph by David Malin
4.5 Max-Planck-Institut für Radioastronomie
4.6 NASA
4.7 © Royal Observatory, Edinburgh & Anglo-Australian Observatory, photograph by David Malin
4.8 R.J. Maddalena, M. Morris, J. Moscowitz & P. Thaddeus, Columbia University
4.9 S.D. Hunter, for the EGRET Collaboration, NASA Goddard Space Flight Center
4.10 Space Telescope Science Institute & NASA/ESA
4.11 Fred Espenak
4.12 R. Bohlin & T. Stecher, NASA Goddard Space Flight Center
4.13 © Anglo-Australian Observatory, photograph by David Malin
4.14 X-ray: W. Hsin-Min Ku, Columbia Astrophysics Laboratory; optical: Lick Observatory photograph
4.15 W. Hsin-Min Ku, Columbia Astrophysics Laboratory
4.16 NRAO/AUI. Observer: F. Yusef-Zadeh
4.17 Peter M. Perry & Barry E. Turnrose, Science Programs, Computer Sciences Corporation
4.18 F.P. Schloerb, Five College Radio Astronomy Observatory; image processing by John C. Good
4.19, 4.20 K. Krisciunas, Royal Observatory Edinburgh
4.21 M. McCaughrean, © Royal Observatory Edinburgh, 1989
4.22 © Royal Observatory, Edinburgh, 1987
4.23 J.F. Arens, G. Lamb & M. Peck, NASA Goddard Space Flight Center, with assistance of W. Hoffman, Steward Observatory, & G. Fazio, Smithsonian Astrophysical Observatory
4.24 John Gleason, Celestial Images
4.25 Max-Planck-Institut für Radioastronomie
4.26 © Infrared Processing and Analysis Center, California Institute of Technology
4.27 Optical data by H. Dickel & T. Gull, using 0.9 m telescope of The Kitt Peak National Observatory; radio data by R. Harten, Westerbork Synthesis Radio Telescope
4.28 Ian Gatley, National Optical Astronomy Observatories

4.29 P. Scott, Mullard Radio Astronomy Observatory, Mon. Not. Roy. Astron. Soc. 194, 23P (1981); image processing by S. Gull & J. Fielden
4.30 Mark Reid: VLBI image produced at NRAO
4.31 © Royal Observatory, Edinburgh, 1992
4.32 © Anglo-Australian Observatory
4.33 X-ray Astronomy Group, Leicester University, & Harvard-Smithsonian Center for Astrophysics
4.34 © Anglo-Australian Observatory, photograph by David Malin
4.35 Courtesy of G. Gehring, R. Rigaut and the COME-ON team
4.36 Space Telescope Science Institute & NASA/ESA
4.37 © Royal Observatory, Edinburgh & Anglo-Australian Observatory, photograph by David Malin
4.38 © Infrared Processing and Analysis Center, California Institute of Technology
4.39 Max-Planck-Institut für Extraterrestrische Physik

5.1 Julian Baum
5.2 © Infrared Processing and Analysis Center, California Institute of Technology
5.3 Ian Gatley, National Optical Astronomy Observatories
5.4 © 1987 Royal Observatory, Edinburgh. Photo by B.W. Hadley
5.5 NASA
5.6 Courtesy of Fokker
5.7 D. Golimowski & S. Durrance, Johns Hopkins University, & M. Clampin, Space Telescope Science Institute
5.8 © 1987 Royal Observatory, Edinburgh. Photo by B.W. Hadley
5.9 © 1992 Royal Observatory, Edinburgh
5.10 European Space Agency
5.11 Mount Stromlo and Siding Spring Observatories, Australian National University

6.1 E. Kallas & W. Reich, based on observations with the Effelsberg 100 m telescope of the Max-Planck-Institut für Radioastronomie & processed with the Astronomical Image Processing System (BABSY) at Bonn University
6.2 © Royal Observatory, Edinburgh, 1989. Photo by B.W. Hadley
6.3 IMB Collaboration
6.4 P. Nisenson & M. Karouska, Harvard-Smithsonian Center for Astrophysics
6.5 Thomas A. Prince, California Institute of Technology
6.6 Radio data: Australia Telescope National Facility; optical data: Space Telescope Science Institute & NASA/ESA
6.7 C. Burrows, ESA/Space Telescope Science Institute
6.8 © Anglo-Australian Observatory, photograph by David Malin
6.9 Palomar Observatory photo, © California Institute of Technology
6.10 Courtesy J. Hester, Arizona State University and Palomar Observatory
6.11 A. Uomoto, Johns Hopkins University, & G.M. MacAlpine, University of Michigan, Astron. J. 93 (6), June 1987
6.12 NRAO/AUI. Observers: A.S. Wilson, D.E. Hogg & N.H. Samarasinha
6.13 M.F. Bietenholz & P.P. Kronberg, University of Toronto, 'Faraday rotation and physical conditions in the Crab Nebula', Astrophys. J. 368: 231–240, 1991 February 10
6.14 NASA
6.15 F.R. Harnden, Jr., Harvard-Smithsonian Center for Astrophysics
6.16 Lick Observatory photographs
6.17 Space Telescope Science Institute & NASA/ESA
6.18 EGRET Collaboration & NASA

6.19 Palomar Observatory photograph by S. van den Bergh, © California Institute of Technology
6.20 Max-Planck-Institut für Extraterrestrische Physik
6.21 S. Gull & G. Pooley, Mullard Radio Astronomy Observatory; D. Green & S. Gull, *IAU Symposium* **101**; image processing by S. Gull & J. Fielden
6.22 Palomar Observatory photo, © California Institute of Technology
6.23 NRAO/AUI. Observers: R.J. Tuffs, R.A. Perley, M.T. Brown & S.F. Gull
6.24 False-colour composite prepared at David Dunlap Observatory, Ontario, by Karl Kamper, from plates obtained by S. van den Bergh with 5 m telescope, Palomar Observatory
6.25 Harvard-Smithsonian Center for Astrophysics
6.26 J.R. Dickel, S.S. Murray, J. Morris & D.C. Wells, first published in *Astrophys. J.* **257**, 145 (1982). Optical data by S. van den Bergh & Dodd, 5 m Hale Telescope, *Astrophys. J.* **162**, 485 (1970); radio data by J.R. Dickel & E.W. Greisen, NRAO interferometer, *Astron. Astrophys.* **75**, 44 (1979); X-ray data by S.S. Murray *et al*, Harvard-Smithsonian Center for *Astrophysics. Astrophys. J. Lett.* **234**, L69 (1979)
6.27 Photolabs, Royal Observatory, Edinburgh; original negative by UK Schmidt Telescope Unit
6.28 S. Bowyer & R. Malina, Center for EUV Astrophysics, University of California, Berkeley
6.29 Max-Planck-Institut für Extraterrestrische Physik
6.30 D.K. Milne, CSIRO, *Austr. J. Phys.* **21**, 203; colouring by David Parker
6.31 B.A. Peterson, P.G. Murdin, P.T. Wallace, R.N. Manchester, Anglo-Australian Observatory/ Starlink
6.32 Craig Markwardt & Hakki Ögelman, University of Wisconsin – Madison
6.33 © Royal Observatory, Edinburgh & Anglo-Australian Observatory, photograph by David Malin
6.34 Palomar Observatory photograph by S. van den Bergh, © California Institute of Technology
6.35 NRAO/AUI. Observers: S.A. Baum & R. Elston
6.36 NRAO/AUI
6.37 David Parker
6.38 X-ray Astronomy Group, Leicester University, & Harvard-Smithsonian Center for Astrophysics
6.39 Palomar Sky Survey, © California Institute of Technology
6.40 Einstein X-ray telescope image of the supernova remnant G109.1.–1.0 (courtesy P.C. Gregory)
6.41 Radio image of the supernova remnant G109.1.–1.0, obtained with the Very Large Array at a wavelength of 21 cm (courtesy P.C. Gregory)

7.1 Julian Baum
7.2 Jerry Mason
7.3 Michael Marten
7.4, 7.5 Kurt W. Weiler, National Science Foundation
7.6 G. Hutschenreiter, Max-Planck-Institut für Radioastronomie, Bonn
7.7 R. Beck & G. Grave, Max-Planck-Institut für Radioastronomie, Bonn; colour radio map produced at computing centre of Rheinisches Landesmuseum, Bonn, by R. Beck
7.8 R.J. Dettmar, Astronomical Institute of the University of Bonn, & R. Beck, Max-Planck-Institut für Radioastronomie, Bonn
7.9 R. Beck, Max-Planck-Institut für Radioastronomie, Bonn
7.10 Seth Shostak
7.11 NRAO/AUI. Observers: J. Dreher & E. Feigelson
7.12 David Parker
7.13 R. Saunders, G. Pooley, J. Baldwin & P. Warner, Mullard Radio Astronomy Observatory, *Mon. Not. Roy. Astron. Soc.* **197**, 287 (1981); image processing by S. Gull & J. Fielden

7.14 Photo: J. Masterson, © CSIRO
7.15 © 1982 Douglas W. Johnson
7.16 © 1989 Royal Observatory, Edinburgh. Photo by B.W. Hadley

8.1 Courtesy of Lund Observatory
8.2 NASA
8.3 G. Haslam et al, Max-Planck-Institut für Radioastronomie (Germany), using observations from Effelsberg (Germany), Jodrell Bank (UK) & Parkes (Australia); colour radio map produced at computing centre of Rheinisches Landesmuseum, Bonn, by R. Beck
8.4 K. Wood, Naval Research Laboratory
8.5 The EGRET Collaboration; map produced by Max-Planck-Institut für Extraterrestrische Physik; courtesy of G. Kanbach
8.6 J.V. Feitzinger, J.A. Stüwe, Astronomical Institute, Ruhr University, Bochum; colour coding by G. Evans, SPL. This picture unifies for the first time the northern and southern dark cloud catalogues of B.T. Lynds, *Astrophys. J. Suppl.* **7**, 1 (1962), and J.V. Feitzinger & J.A. Stüwe, *Astron. Astrophys. Suppl.* **58**, 365 (1984)
8.7 Courtesy Thomas Dame, Harvard-Smithsonian Center for Astrophysics
8.8 NASA
8.9 Carl Heiles, University of California, Berkeley
8.10 Carl Heiles, University of California, Berkeley, & Edward B. Jenkins, Princeton University
8.11 US Naval Observatory
8.12 Max-Planck-Institut für Extraterrestrische Physik
8.13 Thomas A. Prince, California Institute of Technology
8.14 Radio jets from the compact Galactic Centre annihilator found by I.F. Mirabel (CEA-Saclay) *et al.* with the VLA
8.15 © Anglo-Australian Observatory, photograph by David Allen
8.16 D.L. DePoy & N.A. Sharp, National Optical Astronomy Observatories
8.17 NRAO/AUI. Observers: F. Yusef-Zadeh & M.R. Morris
8.18 Radiograph obtained by N. Killeen & K.Y. Lo using the Very Large Array of NRAO, and processed at the NCSA at the University of Illinois
8.19, 8.20 European Southern Observatory
8.21 Max-Planck-Institut für Radioastronomie
8.22 George R. Carruthers, US Naval Research Laboratory
8.23 Max-Planck-Institut für Extraterrestrische Physik
8.24 National Optical Astronomy Observatories
8.25 NASA Goddard Space Flight Center
8.26 © Infrared Processing and Analysis Center, California Institute of Technology
8.27 Courtesy of Sarah Heap, NASA Goddard Space Flight Center

9.1 Julian Baum
9.2, 9.3 NASA
9.4 European Space Agency
9.5 NASA
9.6 David Parker
9.7, 9.8, 9.9 NASA
9.10 Center for EUV Astrophysics, University of California, Berkeley

10.1 © Anglo-Australian Observatory
10.2 Lick Observatory photograph
10.3 Smithsonian Astrophysical Observatory
10.4, 10.5 E. Brinks: 1983, Ph.D. thesis, Leiden Observatory
10.6 © Infrared Processing and Analysis Center, California Institute of Technology

10.7 R. Beck, E.M. Berkhuijsen & R. Wielebinski, Max-Planck-Institut für Radioastronomie, Bonn; colour radio map produced at computing centre of Rheinisches Landesmuseum, Bonn, by R. Beck
10.8, 10.9, 10.10 R. Bohlin & T. Stecher, NASA Goddard Space Flight Center
10.11 L. Van Speybroeck *et al.*, *Astrophys. J. Lett.* **234**, L45
10.12 W.E. Celnik, Astronomische Arbeitsgemeinschaft Bochum (AABO)
10.13 Laboratory for High Energy Astrophysics, NASA Goddard Space Flight Center; courtesy of Frank Primini, Harvard-Smithsonian Center for Astrophysics
10.14 L. Van Speybroeck *et al.*, *Astrophys. J. Lett.* **234**, L45
10.15 J. Lorre, NASA Jet Propulsion Laboratory
10.16 R. Bohlin & T. Stecher, NASA Goddard Space Flight Center
10.17 D.M. Elmegreen, *Astrophys. J. Supp. Series* **47**, 229 (1981). Photograph taken at Palomar Observatory
10.18 Kapteyn Laboratorium, University of Groningen
10.19 W. Hsin-Min Ku, Columbia Astrophysics Laboratory
10.20 Georges Courtes, Observatoire de Marseille & Laboratoire d'Astronomie Spatiale du CNRS (Marseille), Henri Petit, Observatoire de Marseille, & Jean-Pierre Sivan, Laboratoire d'Astronomie Spatiale du CNRS (Marseille).
10.21 NRAO/AUI. Observers: A.H. Rots, J.M. van der Hulst, P.E. Seiden, R.C. Kennicutt, P.C. Crane, A. Bosma, L. Athanassoula & D.M. Elmegreen
10.22 US Naval Observatory
10.23 Kapteyn Laboratorium, University of Groningen
10.24 © Royal Observatory, Edinburgh, 1988
10.25 D. Zaritsky, H.-W. Rix & M. Rieke, 'Inner spiral structure of the galaxy M51', *Nature* **364**, 22 July 1993
10.26 J. Donas, B. Milliard, M. Laget, Laboratoire d'Astronomie Spatiale du CNRS (France). D. Huguenin, Observatoire de Genéve (Switzerland). Image obtained with the FOCA balloon-borne telescope
10.27 Space Telescope Science Institute & NASA/ESA
10.28 J. Donas, B. Milliard, M. Laget, Laboratoire d'Astronomie Spatiale du CNRS (France). D. Huguenin, Observatoire de Geneve (Switzerland). Image obtained with the FOCA balloon-borne telescope
10.29 National Optical Astronomy Observatories
10.30 R. Gräve, Max-Planck-Institut für Radioastronomie, Bonn; colour radio map produced at the computing centre of Rheinisches Landesmuseum, Bonn, by R. Beck
10.31 Max-Planck-Institut für Extraterrestrische Physik
10.32 Colour representation made by R.J. Allen, R. Ekers, J.P. Terlouw & J.M. van der Hulst, Kapteyn Laboratorium (The Netherlands), T.R. Cram & A.H. Rots, NRAO (USA), & the computing & photographic services of the University of Groningen
10.33 Palomar Observatory photo, © California Institute of Technology
10.34 Phil Appleton, Dept. of Astronomy, University of Manchester, & staff of Manchester Starlink node
10.35 M.S. Yun, P.Y.P. Ho & K.Y. Lo, 'A high-resolution image of atomic hydrogen in the M81 group of galaxies', *Nature* **372**, 8 December 1994
10.36 Original photo by H.J. Arp, Mt Wilson Observatory; image processing by J. Lorre, NASA Jet Propulsion Laboratory
10.37 Kapteyn Laboratorium, University of Groningen
10.38, 10.39 NASA
10.40 Max-Planck-Institut für Extraterrestrische Physik

index